Exponents, Roots and Radicals

Herbert M Sauro
Version 1.04
April, 2025

ISBN: 978-1-7325486-5-7

Ambrosius Publishing
Seattle

Contents

Preface

There were many excellent mathematics textbooks published from the early part of the last century that cover both high school and college level.

I have therefore completely re-typeset, rewritten and added additional topics to chapter eight on exponents and radicals from the 1922 public domain copy of "Intermediate algebra for college students" by Peterson, Thurman Stewart. Other than almost a compete replacement of the text to make it more modern, I have added new proofs, exercises and clarification in various parts of the work. In addition, I have added new sections on computing square roots using modern numerical methods, and have also updated some of the exercises to make them more contemporary. Finally, answers to all exercises have been provided rather than just the odd numbered questions that is frequently the case in textbooks. A number of minor errors have been identified by readers such as Guy and Ron and are corrected in this edition. Many thanks to them. I have not added any new copyright notice since the exercises remain the same as in the original publication, though the prose, corrections, additional proofs, new sections, clarifications and answers to exercises are completely new.

All the answers to the exercises have been reviewed in detail at least four times. If, however, you find any errors, please contact me at hsauro@gmail.com and you will receive a free pdf of the book with the new updates.

I would like to dedicate this book to my two sons, Theodore and Tyler who are now just learning about exponents in school.

Finally I would like readers such as

Herbert M. Sauro
April 2023/2025
♣

EXPONENTS, ROOTS, AND RADICALS

Lesson 1: Laws of Exponents

Let's say we have a number given by the symbol a. When we use the notation a^n, we mean that a is multiplied by itself, n times. Let's focus on the case when n is a positive integer since that is the easiest to consider.

Examples: (1) $a^1 = a$,

(2) $a^4 = a \cdot a \cdot a \cdot a$,

(3) $a^n = a \cdot a \cdot \cdots \cdot a$ (n factors).

We will call a^n the nth power of a; where n is called the **exponent**, and a the **base**. There are a number of rules (also called laws) that we can devise for a^n. They include the following:

I. $\boldsymbol{a^m \cdot a^n = a^{m+n}}$.

Proof.

$$\begin{aligned} a^m \cdot a^n &= (a \cdot a \cdot a \cdots\cdots \text{ to } m \text{ facturs })(a \cdot a \cdot a \cdots \text{ to } n \text{ factors }) \\ &= a \cdot a \cdot a \cdots\cdots \text{ to } (m+n) \text{ factors}, \\ &= a^{m+n} \end{aligned}$$

Examples: $x^5 \cdot x^3 = x^8$; $(a+b)^2 \cdot (a+b)^4 = (a+b)^6$.

II. (1) $\boldsymbol{\dfrac{a^m}{a^n} = a^{m-n}}$, if m is greater than n.

(2) $\boldsymbol{\dfrac{a^m}{a^n}} = 1$, if m is equal to n.

(3) $\boldsymbol{\dfrac{a^m}{a^n} = \dfrac{1}{a^{n-m}}}$, if m is less than n.

Proof. (1) $\frac{a^m}{a^n} = \frac{a \cdot a \cdot a \cdots \text{ to } m \text{ factors}}{a \cdot a \cdot a \cdots\cdots \text{ to } n \text{ factors}}$

$$= \frac{a \cdot a \cdot a \cdots \text{ to } (m-n) \text{ factors}}{1},$$

$$= a^{m-n}.$$

We leave the proof of (2) and (3) to the reader.

Examples: (1) $\frac{x^7}{x^3} = x^{7-3} = x^4.$

(2) $\frac{y^5}{y^5} = 1.$

(3) $\frac{x^2}{x^6} = \frac{1}{x^{6-2}} = \frac{1}{x^4}.$

III. $(a^m)^n = a^{mn}$.

Proof. $(a^m)^n = a^m \cdot a^m \cdot a^m \cdots$ to n factors,

$$= a^{m+m+m+\cdots} \text{ to } n \text{ terms},$$

$$= a^{mn}.$$

If is also worth mentioning that the m and n can be swapped without a change to the result, thus:

$$(a^m)^n = (a^n)^m$$

This is a useful trick to know.

Examples: $(x^4)^3 = x^{12}$; $(x^3)^4 = x^{12}$ $(2^3)^2 = 2^6 = 64.$

IV. $(ab)^n = a^n b^n$.

Proof. $(ab)^n = (ab) \cdot (ab) \cdot (ab) \cdots\cdots$ to n factors,

$$= (a \cdot a \cdot a \cdots \text{ on factors })(b \cdot b \cdot b. \quad \text{ to } n \text{ factors }),$$

$$= a^n b^n.$$

Examples:

$$(2x)^3 = 2^3 x^3 = 8x^3; \quad (3x^2 y)^4 = (3)^4 (x^2)^4 (y)^4 = 81x^8 y^4.$$

V. $\left(\dfrac{a}{b}\right)^n = \dfrac{a^n}{b^n}$

Proof.

$$\left(\frac{a}{b}\right)^n = \left(\frac{a}{b}\right)\cdot\left(\frac{a}{b}\right)\cdot\left(\frac{a}{b}\right)\cdots\cdot \text{ to } n \text{ factors,}$$

$$= \frac{a \quad a\cdot a\cdots\cdot \text{ to } n \text{ factors}}{b\cdot b\cdot b\cdots\cdot \text{ to } n \text{ factors}}$$

$$= \frac{a^n}{b^n}.$$

Examples: $\left(\dfrac{x}{3}\right)^2 = \dfrac{x^2}{3^2} = \dfrac{x^2}{9}$; $\left(\dfrac{2x^2}{y^3}\right)^3 = \dfrac{(2)^3\left(x^2\right)^3}{\left(y^3\right)^3} = \dfrac{8x^6}{y^9}$

Be aware that exponents can be attached to collections of terms, such as $(4x)^2$. For example:

(1) $5x^2$ means $5\cdot x\cdot x$, whereas $(5x)^2$ means $5\cdot 5\cdot x\cdot x$.

(2) $-\left(x^3\right)^2$ means $-\left(x^3\right)\left(x^3\right) = -x^6$, whereas $\left(-x^3\right)^2$ means $\left(-x^3\right)\left(-x^3\right) = x^6$.

EXERCISES 1

Apply the laws of exponents to the following problems:

1. $x^3\cdot x^4$ **2.** $m^2\cdot m^6$ **3.** $a\cdot a^2\cdot a^3$

4. $3^2\cdot 3\cdot 3^4$ **5.** $y^8\cdot y^{10}\cdot y^5$ **6.** $(-2)\cdot(-2)^4\cdot(-2)^2$

7. $a^2x^3\cdot a^3x^4$ **8.** $ab^2c\cdot a^2bc^3$ **9.** $x^2y\cdot x^3y\cdot xy^2$

10. $mn\cdot m^2n^3\cdot m^2n$ **11.** $\dfrac{x^5}{x^2}$ **12.** $\dfrac{y^4}{y^6}$

13. $\dfrac{z^7}{z^7}$ **14.** $\dfrac{x^3y^4}{xy^2}$ **15.** $\dfrac{a^3b}{a^2b^2}$

16. $\dfrac{xy^2}{x^3y^3}$ **17.** $\dfrac{a^2b^2}{ab^2}$ **18.** $\dfrac{a^3x^2}{a^3x^2}$

19. $\dfrac{x^3yz^2}{xyz^4}$ **20.** $\dfrac{abc^5}{ab^2c^3}$ **21.** $\left(x^3\right)^6$

22. $(-x)^4$ **23.** $-\left(b^8\right)^2$ **24.** $\left(a^5\right)^5$

25. $\left(-m^2\right)^3$ **26.** $\left(y^6\right)^4$ **27.** $\left(x^2\right)^3\cdot\left(y^3\right)^2$

28. $\left(a^3\right)^5\cdot\left(x^4\right)^4$ **29.** $\left(a^2\right)^2\cdot\left(b^3\right)^3\cdot\left(c^4\right)^4$ **30.** $\left(a^2\right)^4\cdot\left(x^3\right)^3\cdot\left(y^4\right)^2$

31. $(2x^3)^3$ **32.** $(-3a)^2$ **33.** $-(a^2b^3)^4$

34. $(-x^5y^4)^3$ **35.** $(2ab^2)^3$ **36.** $(ax^3)^2$

37. $(a^2b^3c)^4$ **38.** $(x^3y^2z^4)^7$ **39.** $(2ax^2y^3)^4$

40. $(-3m^2nx)^3$ **41.** $\left(\frac{3}{4}\right)^3$ **42.** $\left(-\frac{x}{4}\right)^2$

43. $\left(\frac{a}{x^2}\right)^4$ **44.** $\left(\frac{5}{p}\right)^3$ **45.** $\left(\frac{a^3}{b^5}\right)^5$

46. $\left(\frac{2x}{3y^2}\right)^3$ **47.** $\left(\frac{5a^3}{6b^2}\right)^2$ **48.** $\left(\frac{ab^2}{2x^3}\right)^5$

49. $\left(\frac{ab^2c}{x^2y^3z}\right)^{10}$ **50.** $\left(-\frac{x^2y^3}{2a^2b}\right)^8$

Simplify the following expressions:

51. $(ax^2)^2 \cdot (ax^3)^3$ **52.** $(mn)^2 \cdot (2m^2n)^3$ **53.** $\frac{(x^2y^3)^3}{(x^4y)^4}$

54. $\frac{(b^3c^5)^3}{(b^2c)^2}$ **55.** $(x^3y)^2 \cdot (x^2y^2)^3$ **56.** $(ab^2)^4 \cdot (a^2b)^3$

57. $(2ax^2)^3 \cdot (3bx^3)^2$ **58.** $(5a^2x^3)^2 \cdot (b^3y^4)^3$ **59.** $\frac{(ab^2)^3}{(ab^3)^2}$

60. $\frac{(3p^2q)^3}{(6pq^2)^2}$ **61.** $\frac{(xy^2)^3 \cdot (x^2y^3)^2}{(x^5y)^2}$ **62.** $\frac{(6a^5b^3)^4}{(3ab^2)^3(2a^2b)^5}$

63. $\left(\frac{m^2}{n^3}\right)^3\left(\frac{n}{m^3}\right)^2$ **64.** $\left(\frac{2a^2}{x}\right)^3\left(\frac{x^2}{2a}\right)^4$

65. $\left(\frac{ax^2}{b^2y}\right)^3\left(\frac{by^2}{a^2x}\right)^2$ **66.** $\left(\frac{x^2y}{z^3}\right)^2\left(\frac{x^3z^2}{y^2}\right)^3$

67. $(a^2b^3)^4\left(\frac{x^2}{a^2}\right)^2\left(\frac{y}{b^3}\right)^3$ **68.** $\left(\frac{ax^2}{y}\right)^2\left(\frac{x^2y^3}{a}\right)^3\left(\frac{ay}{x^3}\right)^2$

69. $\left(\frac{2a}{3c}\right)^2\left(\frac{6c}{b^2}\right)^3\left(\frac{b^3}{4a}\right)^2$ **70.** $\left(\frac{a^2b}{x}\right)^4\left(\frac{a}{b^2x^3}\right)^2\left(\frac{x^2}{a^2}\right)^6$

Lesson 2: Roots

Given an expression such as $x = a^n$, it is often the case that we know x and n and we would like to know the value of a. Values for a are called roots. From simple arithmetic we know that:

$$5 \times 5 = 5^2 = 25,$$
$$5 \times 5 \times 5 = 5^3 = 125,$$
$$5 \times 5 \times 5 \times 5 = 5^4 = 625; \text{ and so on.}$$

In other words:

5 is the square root of 25,
5 is the cube root of 125,
5 is the fourth root of 625; and so on.

Given $x = a^n$, we call a the nth root of x. For example, 5 is the square root of 25.

Because finding roots is so common, a special symbol has been devised to indicate the root of a given number. That symbol is $\sqrt{\ }$, and is called the **radical sign**. To indicate which root is meant, a number is placed next to the radical sign. This number is called the index or order of the root. The number under the radical sign is called the **radicand**.

This means, in the expression $\sqrt[3]{8} = 2$, 8 is the radicand, 2 is the root, and 3 is the index. In this case the radical is said to be of the third order.

If the index of a root is not shown, it is implied that it is 2. Thus, $\sqrt{16}$ means $\sqrt[2]{16}$, which equals 4. When the index or order of the root is left out, as in $\sqrt{25}$, the root is the square root. Hence when writing $\sqrt{36}$ we mean the square root of 36.

Since $(+2)^2 = 4$ and $(-2)^2 = 4$, we see that the number 4 has two square roots, namely, +2 and -2. However, we only usually indicate the positive root. Thus, although the square roots of 49 are $+7$ and -7, the expression $\sqrt{49}$ represents only the root 7. Often this value is called the principal root of the radicand.

Example 1. What are the square roots of $64a^2$?

Solution. Since $8a \cdot 8a = 64a^2$, the square roots of $64a^2$ are $\pm 8a$.

Example 2. What is the cube root of $-27a^3b^6$?

Solution. Since $(-3ab^2)^3 = -27a^3b^6$, the cube root of $-27a^3b^6$ is $-3ab^2$.

Example 3. Evaluate $\sqrt[4]{16a^4} + \sqrt[7]{a^7}$.

Solution. Since $(2a)^4 = 16a^4$ and $(a)^7 = a^7$,

$$\sqrt[4]{16a^4} + \sqrt[7]{a^7} = 2a + a = 3a.$$

EXERCISES 2

In the following exercises you may use calculator but you will learn more if you try to solve the problems by hand.

Determine the square roots of the following:

1. 16	**2.** 36	**3.** 64	**4.** 81	**5.** a^2
6. b^4	**7.** x^4y^2	**8.** 49 m^2	**9.** $25a^4$	**10.** $400c^{10}$

Determine the cube root of each of the following:

11. 8	**12.** 27	**13.** −64	**14.** 125	**15.** a^3
16. $-c^9$	**17.** $64x^3y^3$	**18.** 1000 m^6	**19.** $-a^6b^3$	**20.** $\frac{8}{27}a^3$

Determine the roots of the following:

21. $\sqrt{121}$	**22.** $-\sqrt{144}$	**23.** $\sqrt[3]{27}$	**24.** $\sqrt{225}$	**25.** $\sqrt[4]{16}$
26. $\sqrt[3]{729}$	**27.** $\sqrt[3]{-8}$	**28.** $-\sqrt{625}$	**29.** $\sqrt[2]{\frac{1}{64}}$	**30.** $\sqrt[5]{32}$
31. $\sqrt[4]{b^8}$	**32.** $\sqrt[3]{64m^3n^3}$			

Simplify and combine the following:

33. $\sqrt{4} + \sqrt[3]{8}$	**34.** $\sqrt[4]{16} + \sqrt[3]{-27}$	**35.** $\sqrt[3]{64a^3} - \sqrt{a^2}$
36. $\sqrt[5]{a^5} + \sqrt{9a^2}$	**37.** $\sqrt[3]{-8} + \sqrt{144}$	**38.** $\sqrt[4]{81} - \sqrt[8]{1}$
39. $\sqrt{81a^4} - \sqrt[3]{-8a^6}$	**40.** $\sqrt[4]{16a^4b^4} - \sqrt[7]{a^7b^7}$	**41.** $\sqrt[4]{\frac{16}{81}} - \sqrt{\frac{1}{4}}$

Lesson 3: Perfect Squares

This is an easy lesson.

A perfect square is a number made by squaring a whole number. For example, if we square the whole number 3, we get 9. Therefore 9 is a perfect square. 16 is also a perfect square because it can be computed from 4×4.

However, most numbers are not perfect squares. For example, 11 is not a perfect square since there is no whole number when multiplied by itself gives 11.

Here is a table of perfect squares. It's a good idea to memorize the perfect squares that are in the first column of the table.

Table Of Perfect Squares.

You should commit to memory the squares in the first column.

$\sqrt{1}=1$	$\sqrt{121}=11$	$\sqrt{441}=21$	$\sqrt{961}=31$	$\sqrt{1681}=41$
$\sqrt{4}=2$	$\sqrt{144}=12$	$\sqrt{484}=22$	$\sqrt{1024}=32$	$\sqrt{1764}=42$
$\sqrt{9}=3$	$\sqrt{169}=13$	$\sqrt{529}=23$	$\sqrt{1089}=33$	$\sqrt{1849}=43$
$\sqrt{16}=4$	$\sqrt{196}=14$	$\sqrt{576}=24$	$\sqrt{1156}=34$	$\sqrt{1936}=44$
$\sqrt{25}=5$	$\sqrt{225}=15$	$\sqrt{625}=25$	$\sqrt{1225}=35$	$\sqrt{2025}=45$
$\sqrt{36}=6$	$\sqrt{256}=16$	$\sqrt{676}=26$	$\sqrt{1296}=36$	$\sqrt{2116}=46$
$\sqrt{49}=7$	$\sqrt{289}=17$	$\sqrt{729}=27$	$\sqrt{1369}=37$	$\sqrt{2209}=47$
$\sqrt{64}=8$	$\sqrt{324}=18$	$\sqrt{784}=28$	$\sqrt{1444}=38$	$\sqrt{2304}=48$
$\sqrt{81}=9$	$\sqrt{361}=19$	$\sqrt{841}=29$	$\sqrt{1521}=39$	$\sqrt{2401}=49$
$\sqrt{100}=10$	$\sqrt{400}=20$	$\sqrt{900}=30$	$\sqrt{1600}=40$	$\sqrt{2500}=50$

There are also perfect **cubes**, such as 1, 8, 27, 64, etc. The first four of these are worth memorising.

Lesson 4: Fractional Exponents

So far we have assumed that the exponents are positive and whole, for example a^2, or x^7. What happens if we allow fractional exponents such as $a^{\frac{1}{2}}$? What on earth can that mean? One thing we can try is to see what happens if we use the first rule of exponents ($a^n a^m = a^{n+m}$) on a fractional exponent. For example, let's multiple $a^{\frac{1}{2}}$ by itself:

$$a^{\frac{1}{2}} \cdot a^{\frac{1}{2}} = a^{\frac{1}{2}+\frac{1}{2}} = a^1 = a.$$

However, we know that (e.g $\sqrt{9}\sqrt{9} = 9$):

$$\sqrt{a} \cdot \sqrt{a} = a$$

that is:

$$a^{\frac{1}{2}} \cdot a^{\frac{1}{2}} = a = \sqrt{a}\sqrt{a}.$$

This means that an exponent such as $a^{\frac{1}{2}}$, has the same meaning as $\sqrt{a}$.

We can do the same with $a^{\frac{1}{3}}$. For example:

$$a^{\frac{1}{3}} \cdot a^{\frac{1}{3}} \cdot a^{\frac{1}{3}} = a^{\frac{1}{3}+\frac{1}{3}+\frac{1}{3}} = a$$

or

$$\left(a^{\frac{1}{3}}\right)^3 = a$$

That is, $a^{\frac{1}{3}}$ is one of the three equal factors of a or,

$$a^{\frac{1}{3}} = \sqrt[3]{a}.$$

Thus, the fractional exponent $\frac{1}{3}$ means cube root.

In general, for an expression such as $a^{\frac{1}{n}}$,

$$a^{\frac{1}{n}} \cdot a^{\frac{1}{n}} \cdots \text{ to } n \text{ factors} = a^{\frac{1}{n}+\frac{1}{n}+\cdots \text{ to } n \text{ terms}} = a^1 = a.$$

As before, we see that $a^{\frac{1}{n}}$ has the same meaning as $\sqrt[n]{a}$. We have therefore discovered a new rule:

$$a^{\frac{1}{n}} = \sqrt[n]{a}$$

For example, $27^{\frac{1}{3}}$ means $\sqrt[3]{27}$

But we can go further. What might an expression such as $a^{\frac{3}{2}}$ mean? To understand this we need to use the third rule of exponents, which was $(a^m)^n = a^{mn}$. To use this rule we can rewrite $a^{\frac{3}{2}}$ as $a^{3\times\frac{1}{2}}$ where we can set $m = 3$ and $n = 1/2$. Hence, we can write $a^{\frac{3}{2}}$ using the third rule of exponents as:

$$a^{mn} = a^{\frac{3}{2}} = a^{3\times\frac{1}{2}} = \left(a^3\right)^{\frac{1}{2}}.$$

But we know that something raised to the $\frac{1}{2}$ represents a square root, hence it must be the case that:

$$a^{\frac{3}{2}} = \left(a^3\right)^{\frac{1}{2}} = \sqrt{a^3}.$$

Using symbols we would write:

$$a^{\frac{m}{n}} = \left(a^m\right)^{\frac{1}{n}} = \sqrt[n]{a^m}.$$

Another point to make is that because the exponents can be swapped as is $(a^n)^m = (a^m)^n$, we can also state as a new rule that:

Rule for Fractional Exponents

$$a^{\frac{m}{n}} = \left(\sqrt[n]{a}\right)^m = \sqrt[n]{a^m}.$$

Such that, given a fractional exponent such as $a^{\frac{m}{n}}$:

(1) The numerator, m, represents the power.

(2) The denominator, n, represents the order of the root.

Example 1. Evaluate $8^{\frac{2}{3}}$

Solution. From the definition,

$$8^{\frac{2}{3}} = \sqrt[3]{8^2} = \sqrt[3]{64} = 4$$

By the interchange trick, it is also true that:

$$8^{\frac{2}{3}} = (\sqrt[3]{8})^2 = (2)^2 = 4$$

All the rules of exponents can be applied to expressions with fractional exponents, that includes multiplication, division etc. as shown in the following examples.

Example 2. Using the laws of exponents, find

(1) $x^{\frac{1}{2}} \cdot x^{\frac{1}{4}}$ (2) $\dfrac{x^{\frac{2}{3}}}{x^{\frac{1}{3}}}$ (3) $\left(x^{\frac{2}{3}}\right)^6$, (4) $\left(x^6 y^3\right)^{\frac{1}{3}}$, (5) $\left(\frac{x^2}{y^3}\right)^{\frac{1}{2}}$

Solutions.

(1) $x^{\frac{1}{2}} x^{\frac{1}{4}} = x^{\frac{1}{2}+\frac{1}{4}} = x^{\frac{3}{4}}$,

(2) $\dfrac{x^{\frac{2}{3}}}{x^{\frac{1}{3}}} = x^{\frac{2}{3}-\frac{1}{3}} = x^{\frac{1}{3}}$,

(3) $\left(x^{\frac{2}{3}}\right)^6 = x^{\frac{12}{3}} = x^4$,

(4) $\left(x^6 y^3\right)^{\frac{1}{3}} = \left(x^6\right)^{\frac{1}{3}} \left(y^3\right)^{\frac{1}{3}} = x^2 y$,

(5) $\left(\dfrac{x^{\frac{1}{3}}}{y^2}\right)^{\frac{1}{2}} = \dfrac{\left(x^{\frac{1}{3}}\right)^{\frac{1}{2}}}{(y^2)^{\frac{1}{2}}} = \dfrac{x^{\frac{1}{6}}}{y}$.

EXERCISES 3

Express the following using the radical form, for example $x^{\frac{1}{2}} = \sqrt{x}$

1. $x^{\frac{1}{2}}, x^{\frac{1}{3}}, y^{\frac{2}{3}}, y^{\frac{3}{4}}, a^{\frac{1}{7}}, a^{\frac{3}{5}}, b^{0.5}, b^{0.4}$

2. $x^{\frac{3}{2}}, x^{1\frac{1}{3}}, y^{\frac{5}{6}}, y^{\frac{6}{5}}, a^{2\frac{1}{2}}, a^{\frac{2}{5}}, (a+b)^{\frac{1}{2}}, (a-b)^{\frac{1}{3}}$

3. $3x^{\frac{1}{2}}, (3x)^{\frac{1}{2}}, 8y^{\frac{1}{3}}, 5y^{\frac{3}{5}}, 2a^{\frac{4}{3}}, 2a^{\frac{3}{4}}, \left(a^2+b^2\right)^{\frac{1}{2}}, \left(a^3-b^3\right)^{\frac{1}{3}}$

4. $4x^{\frac{4}{3}}, 2x^{3\frac{1}{2}}, (4y)^{\frac{2}{3}}, 4y^{\frac{2}{3}}, 3a^{1\frac{1}{2}}, (2a)^{\frac{4}{3}}, 7(a+b)^{\frac{1}{2}}, [7(a+b)]^{\frac{1}{2}}$

Now do the opposite, convert the radical form into the equivalent exponential form:

5. $\sqrt{a}, \sqrt[3]{a}, \sqrt[3]{b^2}, \sqrt{b^3}, \sqrt[5]{x^2}, \sqrt{x^5}, \sqrt[4]{y^3}, \sqrt[3]{y^5}$

6. $\sqrt[3]{a^4}, \sqrt[4]{a}, \sqrt[6]{b^2}, \sqrt[3]{b^7}, \sqrt{x^9}, \sqrt[4]{x^7}, \sqrt{x+y}, \sqrt{x^2+y^2}$

7. $2\sqrt{a}, \sqrt{2a}, 7\sqrt[3]{b^2}, 3\sqrt[5]{b^3}, \sqrt[4]{16x^3}, 5\sqrt{x^5}, 2\sqrt{x-y}, 4\sqrt{x^2-y^2}$

8. $a\sqrt{b}, \sqrt{ab}, b\sqrt{a}, a\sqrt[3]{b^2}, \sqrt[3]{ab^2}, b\sqrt[3]{a}, x\sqrt{x+y}, x\sqrt[3]{x-y}$

Determine the values of each of the following:

9. $25^{\frac{1}{2}}$	**10.** $8^{\frac{1}{3}}$	**11.** $16^{\frac{1}{4}}$	**12.** $27^{\frac{2}{3}}$	**13.** $4^{\frac{3}{2}}$
14. $(-8)^{\frac{2}{3}}$	**15.** $144^{\frac{1}{2}}$	**16.** $32^{\frac{1}{5}}$	**17.** $(-1)^{\frac{3}{5}}$	**18.** $1^{\frac{4}{7}}$
19. $9^{\frac{3}{2}}$	**20.** $81^{\frac{3}{4}}$	**21.** $\left(\frac{1}{8}\right)^{\frac{1}{3}}$	**22.** $\left(\frac{25}{36}\right)^{\frac{1}{2}}$	**23.** $\left(\frac{8}{27}\right)^{\frac{2}{3}}$
24. $\left(\frac{4}{9}\right)^{\frac{3}{2}}$	**25.** $(0.25)^{\frac{1}{2}}$	**26.** $(1.21)^{\frac{1}{2}}$	**27.** $(0.027)^{\frac{1}{3}}$	**28.** $(0.125)^{\frac{2}{3}}$

Using the multiplication law of exponents evaluate the following:

29. $a^{\frac{1}{2}} \cdot a^{\frac{1}{4}}$	**30.** $x^{\frac{1}{3}} \cdot x^{\frac{1}{3}}$	**31.** $b \cdot b^{\frac{1}{2}}$	**32.** $y^{\frac{1}{3}} \cdot y^{\frac{2}{3}}$
33. $2^{\frac{1}{2}} \cdot 2^{\frac{1}{3}}$	**34.** $x^{\frac{3}{2}} \cdot x^{\frac{2}{3}}$	**35.** $n^{\frac{1}{2}} \cdot n^{\frac{1}{4}} \cdot n^{\frac{1}{4}}$	**36.** $a^{\frac{1}{3}} \cdot a^{\frac{2}{3}} \cdot a$

Using the division law of exponents evaluate the following:

37. $a^{\frac{3}{4}} \div a^{\frac{1}{4}}$	**38.** $x^{\frac{5}{6}} \div x^{\frac{1}{3}}$	**39.** $y^{\frac{2}{3}} \div y^{\frac{1}{4}}$	**40.** $2^{\frac{5}{4}} \div 2^{\frac{1}{4}}$
41. $\frac{n^{\frac{4}{3}}}{n^{\frac{3}{4}}}$	**42.** $\frac{a^{0.5}}{a^{0.25}}$	**43.** $\frac{x^{\frac{1}{2}}}{x^{\frac{1}{3}}}$	**44.** $\frac{m^{2\frac{1}{2}}}{m^{1\frac{1}{3}}}$

Simplify the following expressions:

45. $\left(a^{\frac{2}{3}}\right)^6$ **46.** $\left(a^6\right)^{\frac{2}{3}}$ **47.** $\left(x^{\frac{2}{3}}\right)^{\frac{1}{2}}$ **48.** $\left(y^{\frac{2}{3}}\right)^{\frac{3}{2}}$

49. $\left(4^{\frac{3}{4}}\right)^{\frac{2}{3}}$ **50.** $\left(z^{10}\right)^{\frac{2}{5}}$ **51.** $\left(x^{\frac{1}{2}}y^{\frac{1}{3}}\right)^6$ **52.** $\left(a^2b^3\right)^{\frac{1}{6}}$

53. $\left(\frac{x^2}{y^3}\right)^{\frac{5}{6}}$ **54.** $\left(\frac{a^{\frac{3}{2}}}{b}\right)^4$ **55.** $\left(\frac{b^{\frac{3}{4}}}{c^{\frac{2}{3}}}\right)^6$ **56.** $\left(\frac{x^{\frac{1}{2}}}{y^2}\right)^3$

57. $\left(\frac{x^{\frac{1}{2}}\cdot x^{\frac{1}{3}}}{x^{\frac{5}{12}}}\right)^6$ **58.** $\frac{\left(ax^2\right)^{\frac{1}{2}}\cdot\left(a^2x\right)^{\frac{1}{4}}}{x^{\frac{1}{4}}}$ **59.** $\frac{\left(n^{\frac{2}{3}}\right)^2\cdot\left(n^{\frac{3}{5}}\right)^{10}}{\left(n^{\frac{4}{3}}\right)^3}$ **60.** $\frac{\left(y^{\frac{3}{2}}\right)^{\frac{1}{2}}\cdot\left(y^3\right)^{\frac{1}{6}}}{\left(y^{\frac{5}{2}}\right)^{\frac{1}{2}}}$

61. $\left(\frac{a^{\frac{1}{2}}}{b^2}\right)^2\cdot\left(\frac{b^{10}}{a}\right)^{\frac{1}{2}}$ **62.** $\frac{a^{\frac{1}{2}}x^{\frac{2}{3}}\cdot a^{\frac{1}{3}}x^{\frac{1}{4}}}{a^{\frac{2}{3}}x^{\frac{5}{12}}}$

Lesson 5: Zero and Negative Exponents

We next come to look at two odd looking exponents, a^0 and negative exponents such as a^{-2}. What do these mean? Let's look at the a^0 expression first. Whatever a^0 is, it should obey the laws of exponents. For example, I think we would all agree that:

$$\frac{a^3}{a^3} = 1$$

We also can express this using the division rule of exponents (rule II):

$$\frac{a^3}{a^3} = a^{3-3} = a^0 = 1$$

In other words, as odd as it may seem, $a^0 = 1$. Hence, we define:

$$\boldsymbol{a^0 = 1, \quad \text{if } a \neq 0}$$

Note that it imposes the condition that $a \neq 0$, that is a cannot be zero. If it were we'd get the expression 0^0 doesn't makes much sense. So let's avoid that.

Examples: $(-5)^0 = 1, \quad \left(\frac{2}{3}\right)^0 = 1, \quad \left(3x^2\right)^0 = 1.$

Anything raised to 0 will always be equal to one.

The next situation is when the exponent is negative, such as a^{-2}. We can easily figure out what this means by also using the division rule of exponents:

$$a^m \cdot a^{-m} = a^{m-m} = a^0 = 1$$

If we divide both sides by a^m we get the new rule:

$$\boldsymbol{a^{-m} = \frac{1}{a^m} \quad \text{if } a \neq 0}$$

That is a^{-m} is just the reciprocal of a^m.

Examples: $2^{-3} = \dfrac{1}{2^3} = \dfrac{1}{8}$, $9^{-\frac{1}{2}} = \dfrac{1}{9^{\frac{1}{2}}} = \dfrac{1}{3}$, $x^{-1} = \dfrac{1}{x}$.

We have thus discovered two new rules related to a^0 and a^{-m}. An interesting and useful additional rule follows from these which is that numerators and denominators can be swapped simply by changing their sign. For example:

$$\frac{a^2}{b^{-3}} = \frac{b^3}{a^{-2}}$$

This even applies to the simple case:

$$\frac{a}{1} = \frac{1}{a^{-1}}$$

Note that the 1 in the denominator doesn't change sign as you can consider it to be 1^0. Here is a proof:

$$\frac{ax^{-n}}{b} = \frac{ax^{-n}x^n}{bx^n} = \frac{ax^0}{bx^n} = \frac{a}{bx^n},$$

or

$$\frac{a}{bx^{-n}} = \frac{ax^n}{bx^{-n}x^n} = \frac{ax^n}{bx^0} = \frac{ax^n}{b}.$$

Example: $\dfrac{x^{-2}y^3}{a^{-1}b^{-4}} = \dfrac{ax^{-2}}{b^{-4}y^{-3}} = \dfrac{ab^4}{x^2y^{-3}} = \dfrac{ab^4y^3}{x^2}$.

A Trap: There is one trap you might fall into however if you're not careful. The swaps work when you have a product of terms but not a sum. For example, the following is perfectly valid:

$$\frac{1}{a^{-1}b^{-1}} \text{ becomes } \frac{ab}{1} = ab$$

But the following is most certainly not valid:

$$\frac{1}{a^{-1}+b^{-1}} \neq \frac{a+b}{1}$$

Instead

$$\frac{1}{a^{-1}+b^{-1}} \text{ becomes } \frac{1}{\frac{1}{a}+\frac{1}{b}} \text{ or } \frac{ab}{a+b}$$

Example 1. Express $\left(\frac{ax^{-2}}{y^3}\right)^{-2}$ in the simplest form with positive exponents. It's asking us to get rid of the two -2 exponents.

Solution. We swap the outer exponent first, then the inner exponents.

$$\left(\frac{ax^{-2}}{y^3}\right)^{-2}\left(\frac{y^3}{ax^{-2}}\right)^2=\frac{y^6}{a^2x^{-4}}=\frac{a^{-2}x^4}{y^{-6}}=\frac{x^4y^6}{a^2}$$

EXERCISES 4

Find the numerical values of the following:

1. 4^{-1}
2. 3^0
3. 2^{-3}
4. 10^{-2}
5. $(-1)^0$
6. $\left(\frac{2}{3}\right)^{-2}$
7. $\frac{1}{5^{-1}}$
8. $\left(7^{-2}\right)^0$
9. $\left(\frac{1}{2}\right)^{-2}\cdot 4^{-3}$
10. $\left(\frac{2}{5}\right)^{-3}\cdot\left(\frac{3}{4}\right)^0$
11. $8^{-\frac{1}{3}}$
12. 3×2^0
13. $\frac{3^{-2}}{2^{-3}}$
14. $4^{-\frac{1}{2}}\cdot\left(\frac{1}{2}\right)^{-4}$
15. 3×10^{-4}
16. $4^{-\frac{3}{2}}\div 8^{-\frac{2}{3}}$
17. $(0.25)^{-\frac{1}{2}}\cdot\left(-\frac{1}{2}\right)^0$
18. $\left(\frac{2}{3}\right)^{-5}\cdot\left(\frac{3}{4}\right)^{-3}$
19. $\sqrt[3]{-8^{-2}}$
20. $\sqrt[6]{64^{-1}}\cdot\sqrt[5]{10^0}$

Simplify the following expressions using the laws of exponents:

21. $x^{-2}\cdot x^3$
22. $a^2\cdot a^{-\frac{1}{2}}$
23. $m^{-1}\cdot m^{-3}$
24. $b^2\div b^{-2}$
25. $x^0\div x^{-4}$
26. $y^3\div y^{-4}$
27. $\left(x^{-2}\right)^{-3}$
28. $\left(a^{-2}\right)^4$
29. $\left(y^{-\frac{1}{2}}\right)^{-6}$
30. $\left(x^{-2}y^{-4}\right)^{-1}$
31. $\left(a^{-\frac{1}{2}}b^2\right)^{-2}$
32. $\left(a^3x^{-1}\right)^{-4}$

Express the following in the simplest form with positive exponents:

33. a^{-4}
34. $a^{-2}b^3$
35. $\frac{x^{-2}}{5}$
36. $3y^{-3}$
37. $\frac{a^{-3}}{x^{-2}}$
38. $\frac{a^{-1}b^{-2}}{c^{-3}}$
39. $\frac{2x^{-4}}{3y^{-1}}$
40. $\frac{a^2b^{-3}}{x^2y^{-3}}$
41. $x^{-1}-y^{-1}$
42. $\frac{a}{b^{-1}}+\frac{b}{a^{-1}}$
43. $(x-y)^{-1}$
44. $\frac{4^{-1}m^{-2}n^{-3}}{(2mn)^{-5}}$

45. $\frac{x}{y}+\frac{y^{-1}}{x^{-1}}$ **46.** $\frac{x^{-1}+y^{-1}}{z^{-1}}$ **47.** $ab^{-1}+a^{-1}b$

48. $\sqrt{(x+y)^{-2}}$ **49.** $\left(m^{-1}+n^{-1}\right)\left(m^{-1}-n^{-1}\right)$

50. $a(a+b)^{-1}+b(a-b)^{-1}$ **51.** $\frac{x^2-y^2}{x^{-1}+y^{-1}}$

52. $\frac{(a+b)^{-1}}{a^{-1}+b^{-1}}$

Simplify the following expressions:

53. $a^n \cdot a^{3-n}$ **54.** $x^m \cdot x^{m-2}$ **55.** $\frac{b^{n+1}}{b^{2n-1}}$

56. $\frac{y^{-n-2}}{y^{-5}}$ **57.** $\left(\frac{x^n}{y^3}\right)^{n-1}$ **58.** $\left(\frac{a^{3r}}{b^r}\right)^{\frac{1}{r}}$

59. $\left(a^2b^n\right)^n$ **60.** $\left(x^{-a}y^{1-a}\right)^{-2}$ **61.** $\left(a^kb^2\right)^3 \cdot \left(a^{k-1}b^3\right)^{-2}$

62. $x^{n-1}y^{1-n} \cdot x^{2-n}y^{\frac{n}{2}}$ **63.** $\frac{\left(p^rq^{r-1}\right)^r}{\left(p^{r-1}q^r\right)^r}$ **64.** $\frac{m^{x+2}n^{2x-1}}{\left(m^xn^{1-x}\right)^2}$

Lesson 6: Rational and Irrational Numbers

Here is a brief lesson on rational and irrational numbers. A rational number is a number which can be expressed as the ratio of two (positive or negative) integers. Any real number which is not rational is called an irrational number.

Examples:

The numbers $2\frac{2}{3}$, -5, 2.17 can be written as $\frac{8}{3}$, $-\frac{-5}{1}$, $\frac{217}{100}$ and hence are rational numbers.

The number represented by π, $\sqrt{2}$, $\sqrt[3]{5}$ are examples of irrational numbers. We won't provide a proof here but for those interested the proof that $\sqrt{2}$ is irrational was given by Euclid over 2000 years ago. The proof can be found on the internet in many places, for example at https://www.mathsisfun.com/numbers/euclid-square-root-2-irrational.html

An irrational number which is an indicated root of a rational number is called a **surd**.

Thus, $\sqrt{2}$, $\sqrt[3]{5}$ are surds. $\sqrt[3]{8}$ is not a surd because $\sqrt[3]{8}$ can be simplify to 2 which is a rational number. $\sqrt{1+\sqrt{2}}$ is not a surd because $1+\sqrt{2}$ is not a rational number.

Here are a few additional definitions:

A surd of the second order is called a quadratic surd. A binomial which contains at least one **quadratic surd** is called a binomial quadratic surd.

Thus, $2-\sqrt{5}$ and $2\sqrt{3}+\sqrt{7}$ are **binomial quadratic surds.**

Lesson 7: Multiplication of Radicals

Here is a very useful multiplication rule for radicals. Let's give the rule first then we'll show it's true.

$$\boldsymbol{\sqrt[n]{ab} = \sqrt[n]{a} \cdot \sqrt[n]{b}}$$

That is, if a radical has two or more terms, the terms can be unpacked into two or more separate radicals of the same order. The rule is very useful for simplifying and dissembling a radical. For example:

$$\sqrt{4 \cdot 6} = \sqrt{4} \cdot \sqrt{6} = 2\sqrt{6}$$

or equally important is the converse where radicals can be combined:

$$\sqrt{4} \cdot \sqrt{3} = \sqrt{4 \cdot 3} = \sqrt{12}$$

Now let's prove it. From the definition of radicals,

$$\sqrt{p^2 q^2} = \sqrt{(pq)^2} = (pq) = p \cdot q = \sqrt{p^2} \cdot \sqrt{q^2}.$$

If we let $a = p^2$ and $b = q^2$, this result reduces to

$$\sqrt{ab} = \sqrt{a} \cdot \sqrt{b}$$

In general, for roots of any order,

$$\sqrt[n]{p^n q^n} = \sqrt[n]{(pq)^n} = p \cdot q = \sqrt[n]{p^n} \cdot \sqrt[n]{q^n}$$

Again, letting $a = p^n$ and $b = q^n$, we have

$$\sqrt[n]{ab} = \sqrt[n]{a} \cdot \sqrt[n]{b}$$

Another way to look at it which is probably simpler, is to convert to exponential form first:

$$\sqrt[n]{ab} = (ab)^{\frac{1}{n}} = a^{\frac{1}{n}} b^{\frac{1}{n}}$$

But

$$a^{\frac{1}{n}} b^{\frac{1}{n}} = \sqrt[n]{a}\sqrt[n]{b}$$

Hence:

$$\sqrt[n]{ab} = \sqrt[n]{a}\sqrt[n]{b}$$

This can be extended to any number of factors:

$$\sqrt[n]{abc\ldots} = \sqrt[n]{a}\sqrt[n]{b}\sqrt[n]{c}\ldots$$

Example 1. Simplify the radicals $\sqrt{18}, \sqrt[3]{-16}, \sqrt{8a^2b^5}$.

Solution. By the above multiplication rule,

$$\sqrt{18} = \sqrt{9 \cdot 2} = \sqrt{9} \cdot \sqrt{2} = 3\sqrt{2},$$
$$\sqrt[3]{-16} = \sqrt[3]{-8 \cdot 2} = \sqrt[3]{-8} \cdot \sqrt[3]{2} = -2\sqrt[3]{2},$$
$$\sqrt{8a^2b^5} = \sqrt{4a^2b^4 \cdot 2b} = \sqrt{4a^2b^4} \cdot \sqrt{2b} = 2ab^2\sqrt{2b}.$$

Example 2. Multiply and simplify $2\sqrt{6} \cdot 3\sqrt{10}$.

Solution. First multiply to obtain:

$$2\sqrt{6} \cdot 3\sqrt{10} = 2 \cdot 3 \cdot \sqrt{6} \cdot \sqrt{10} = 6 \cdot \sqrt{60}.$$

Next we simplify to obtain:

$$6 \cdot \sqrt{60} = 6 \cdot \sqrt{4 \cdot 15} = 6 \cdot \sqrt{4} \cdot \sqrt{15} = 6 \cdot 2 \cdot \sqrt{15} = 12\sqrt{15}.$$

Example 3. Multiply and simplify $\sqrt{65} \cdot \sqrt{91}$.

Solution. When two radicands of a product contain a common factor (13 in this case), the simplification can be made easier, as follows:

$$\sqrt{65} \cdot \sqrt{91} = \sqrt{13 \cdot 5} \cdot \sqrt{13 \cdot 7} = \sqrt{13} \cdot \sqrt{5} \cdot \sqrt{13} \cdot \sqrt{7}$$
$$= \sqrt{13}\sqrt{13} \cdot \sqrt{5} \cdot \sqrt{7} = 13\sqrt{35}.$$

EXERCISES 5

Multiply and evaluate:

1. $\sqrt{2}\cdot\sqrt{8}$ **2.** $\sqrt{3}\cdot\sqrt{12}$ **3.** $\sqrt{5}\cdot\sqrt{5}$ **4.** $\sqrt[3]{9}\cdot\sqrt[3]{3}$

5. $\sqrt[3]{4}\cdot\sqrt[3]{-2}$ **6.** $\sqrt[4]{8}\cdot\sqrt[4]{2}$ **7.** $\sqrt{a}\cdot\sqrt{4a}$ **8.** $\sqrt{\frac{1}{2}x^3}\cdot\sqrt{18x}$

9. $\sqrt[3]{9b^4}\cdot\sqrt[3]{-3b^2}$

Simplify each of the following radicals:

10. $\sqrt{12}$ **11.** $\sqrt{50}$ **12.** $\sqrt{72}$ **13.** $\sqrt{242}$

14. $\sqrt{75}$ **15.** $\sqrt{162}$ **16.** $\sqrt{147}$ **17.** $\sqrt{800}$

18. $\sqrt[3]{24}$ **19.** $\sqrt[3]{54}$ **20.** $\sqrt[3]{-40}$ **21.** $\sqrt[3]{-250}$

22. $\sqrt[4]{32}$ **23.** $\sqrt[4]{48}$ **24.** $\sqrt{20x^3}$ **25.** $\sqrt{45a^3b^2}$

26. $\sqrt{175\,xy^2}$ **27.** $\sqrt{63a^2}$ **28.** $\sqrt[3]{48x^5}$ **29.** $\sqrt[3]{81\,a^5b^9}$

Multiply and simplify:

30. $\sqrt{2}\cdot\sqrt{10}$ **31.** $\sqrt{7}\cdot\sqrt{21}$ **32.** $\sqrt{6}\cdot\sqrt{15}$ **33.** $\sqrt{35}\cdot\sqrt{45}$

34. $\sqrt{42}\cdot\sqrt{77}$ **35.** $\sqrt{30}\cdot\sqrt{35}$ **36.** $\sqrt[3]{3}\cdot\sqrt[3]{18}$ **37.** $\sqrt[3]{4}\cdot\sqrt[3]{14}$

38. $3\sqrt{10}\cdot 2\sqrt{15}$ **39.** $(2a\sqrt{5a})^2$ **40.** $4\sqrt[3]{6}\cdot 7\sqrt[3]{12}$

Lesson 8: Division of Radicals

The next rule for radicals is the division rule. The rule is quite simple:

$$\sqrt{\frac{\boldsymbol{a}}{\boldsymbol{b}}} = \frac{\sqrt{\boldsymbol{a}}}{\sqrt{\boldsymbol{b}}} \text{ and conversely } \frac{\sqrt{\boldsymbol{a}}}{\sqrt{\boldsymbol{b}}} = \sqrt{\frac{\boldsymbol{a}}{\boldsymbol{b}}}$$

The proof is as follows:

$$\sqrt{\frac{p^2}{q^2}} = \sqrt{\left(\frac{p}{q}\right)^2} = \frac{p}{q} = \frac{\sqrt{p^2}}{\sqrt{q^2}}$$

Hence, letting $a = p^2$ and $b = q^2$, we have

$$\sqrt{\frac{\boldsymbol{a}}{\boldsymbol{b}}} = \frac{\sqrt{\boldsymbol{a}}}{\sqrt{\boldsymbol{b}}} \text{ and } \frac{\sqrt{\boldsymbol{a}}}{\sqrt{\boldsymbol{b}}} = \sqrt{\frac{\boldsymbol{a}}{\boldsymbol{b}}}$$

As before we can also use the simpler argument using the exponent rules as follows:

$$\sqrt[n]{\frac{p}{q}} = \left(\frac{p}{q}\right)^{\frac{1}{n}} = \frac{p^{\frac{1}{n}}}{q^{\frac{1}{n}}} = \frac{\sqrt[n]{p}}{\sqrt[n]{q}}$$

This relationship is true for roots of any order.

Example 1. Simplify $\sqrt{\frac{4}{9}}$.

Solution.

$$\sqrt{\frac{4}{9}} = \frac{\sqrt{4}}{\sqrt{9}} = \frac{2}{3}$$

Example 2. Simplify $\sqrt{\frac{2}{3}}$.

Solution. At first sight there doesn't appear to be a way to simplify this. However, we can use a trick where we multiply the numerator and denominator by the smallest number which will make the denominator a "square number." In this case we multiply by 3, thus,

$$\sqrt{\frac{2}{3}} = \sqrt{\frac{6}{9}} = \frac{\sqrt{6}}{\sqrt{9}} = \frac{\sqrt{6}}{3} = \frac{1}{3}\sqrt{6}.$$

Example 3. Simplify $\frac{2}{3}\sqrt[3]{6\frac{3}{4}}$.

Solution. For cube roots, we try to make the denominator a perfect cube (8 in this case); hence:

$$\frac{2}{3}\sqrt[3]{\frac{27}{4}} = \frac{2}{3}\sqrt[3]{\frac{27\cdot 2}{8}} = \frac{2}{3}\cdot\frac{\sqrt[3]{27}\sqrt[3]{2}}{\sqrt[3]{8}} = \frac{2}{3}\cdot\frac{3\sqrt[3]{2}}{2} = \sqrt[3]{2}$$

Example 3. Simplify $\dfrac{6ab^2}{\sqrt{12ab^3}}$

Solution. Multiplying numerator and denominator by $\sqrt{3ab}$ gives

$$\frac{6ab^2\cdot\sqrt{3ab}}{\sqrt{12ab^3}\cdot\sqrt{3ab}} = \frac{6ab^2\sqrt{3ab}}{\sqrt{36a^2b^4}} = \frac{6ab^2\sqrt{3ab}}{6ab^2} = \sqrt{3ab}$$

EXERCISES 6

Divide and evaluate the following expressions:

1. $\dfrac{\sqrt{12}}{\sqrt{3}}$ **2.** $\dfrac{\sqrt{72}}{\sqrt{2}}$ **3.** $\dfrac{\sqrt{294}}{\sqrt{6}}$ **4.** $\dfrac{\sqrt{180}}{\sqrt{5}}$

5. $\dfrac{\sqrt[3]{40}}{\sqrt[3]{5}}$ **6.** $\dfrac{\sqrt[3]{108}}{\sqrt[3]{4}}$ **7.** $\dfrac{\sqrt[3]{-56}}{\sqrt[3]{7}}$ **8.** $\dfrac{\sqrt[4]{48}}{\sqrt[4]{3}}$

9. $\dfrac{\sqrt{8a^3b}}{\sqrt{2ab}}$ **10.** $\dfrac{\sqrt{75x^7y^8}}{\sqrt{3x^3y^2}}$ **11.** $\dfrac{\sqrt[3]{x^8}}{\sqrt[3]{x^2}}$ **12.** $\dfrac{\sqrt[4]{a^9b^6}}{\sqrt[4]{ab^2}}$

Simplify:

13. $\sqrt{\frac{3}{5}}$ **14.** $\sqrt{\frac{5}{7}}$ **15.** $\sqrt{\frac{5}{12}}$ **16.** $\sqrt{\frac{7}{18}}$

17. $\sqrt{2\frac{1}{2}}$ **18.** $\sqrt{3\frac{5}{9}}$ **19.** $\sqrt{5\frac{2}{5}}$ **20.** $\sqrt{8\frac{1}{6}}$

21. $\sqrt[3]{\frac{3}{4}}$ **22.** $\sqrt[3]{7\frac{1}{9}}$ **23.** $\sqrt{\frac{3a}{2b^3}}$ **24.** $\sqrt{\frac{28}{45n}}$

25. $\sqrt[3]{\frac{x}{y}}$ **26.** $\sqrt[3]{\frac{2a^2}{3b^2}}$ **27.** $\frac{2a}{b}\sqrt{\frac{b^3}{2a}}$ **28.** $\frac{12}{x}\sqrt[3]{\frac{x^3}{12}}$

Divide and simplify the following:

29. $\dfrac{\sqrt{7}}{\sqrt{10}}$ **30.** $\dfrac{\sqrt{75}}{\sqrt{8}}$ **31.** $\dfrac{\sqrt{28}}{\sqrt{5}}$ **32.** $\dfrac{3}{\sqrt{12}}$

33. $\dfrac{2\sqrt{3}}{5\sqrt{20}}$ **34.** $\dfrac{3\sqrt{5}}{2\sqrt{18}}$ **35.** $\dfrac{3\sqrt{8}}{4\sqrt{27}}$ **36.** $\dfrac{5\sqrt{32}}{4\sqrt{45}}$

37. $\dfrac{3\sqrt[3]{5}}{2\sqrt[3]{9}}$ **38.** $\dfrac{5\sqrt[3]{15}}{3\sqrt[3]{5}}$ **39.** $\dfrac{2\sqrt[3]{54}}{3\sqrt[3]{4}}$ **40.** $\dfrac{6y\sqrt{2x^3y}}{5x\sqrt{8xy^3}}$

41. $\dfrac{3ab\sqrt{20c^2}}{2c\sqrt{9a^2b}}$ **42.** $\dfrac{a\sqrt[3]{5a}}{3\sqrt[3]{4a^2}}$ **43.** $\dfrac{2x\sqrt{9xy}}{3\sqrt{8x}}$ **44.** $\dfrac{\sqrt{ax-ay}}{\sqrt{bx-by}}$

Lesson 9: Addition and Subtraction of Radicals

Probably the easiest operation with radicals is adding and subtracting. There are however two restrictions. The first is that only radicals of the same order can be added or subtracted. For example:

$$\sqrt{2}+\sqrt{2}=2\sqrt{2}$$

$$3\sqrt{2}+4\sqrt{2}=7\sqrt{2}$$

while: $\sqrt[3]{4}+\sqrt{4}$ cannot be added together unless we can some how change the orders to be the same (see lesson 12).

The other thing we need to make sure when adding or subtracting radicals is that the radicand is also the same.

For example. $5\sqrt{2}$, $\frac{1}{2}\sqrt{2}$, and $x\sqrt{2}$ can be added because the radicand, 2, and order is the same.

But $\sqrt{2}$, $\sqrt{3}$, and $\sqrt[3]{2}$ cannot be added because the radicand and order are different.

> Radicals that have the same order and radicand can be added or subtracted.

Here are some more examples:

$$15\sqrt{5}+2\sqrt{5}=17\sqrt{5}$$

$$8\sqrt[3]{3}-7\sqrt[3]{3}=\sqrt[3]{3}.$$

Unlike radicals cannot be combined by addition or subtraction unless it is possible to reduce them to like radicals.

Example 1. Combine $\sqrt{18}+\sqrt{72}-\sqrt{32}$.

Solution. Simplifying each radical before combining, we have

$$\sqrt{18}+\sqrt{72}-\sqrt{32}=3\sqrt{2}+6\sqrt{2}-4\sqrt{2}=5\sqrt{2}.$$

Example 2. Combine $\sqrt[3]{16}+\sqrt[3]{61}-\sqrt[3]{54}$.

Solution.

$$\sqrt[3]{16}+\sqrt[3]{64}-\sqrt[3]{54}=2\sqrt[3]{2}+4-3\sqrt[3]{2}=4-\sqrt[3]{2}$$

Example 3. Combine $\sqrt{12}-\frac{1}{\sqrt{2}}-\sqrt{\frac{4}{3}}+\sqrt{3\frac{1}{8}}$.

Solution.

$$\begin{aligned}\sqrt{12}-\frac{1}{\sqrt{2}}-\sqrt{\frac{4}{3}}+\sqrt{3\frac{1}{8}} &= \sqrt{4\cdot 3}-\frac{\sqrt{2}}{\sqrt{2}\sqrt{2}}-\sqrt{\frac{4}{9}\cdot 3}+\sqrt{\frac{25}{16}\cdot 2}, \\ &= 2\sqrt{3}-\frac{1}{2}\sqrt{2}-\frac{2}{3}\sqrt{3}+\frac{5}{4}\sqrt{2} \\ &= \frac{4}{3}\sqrt{3}+\frac{3}{4}\sqrt{2}\end{aligned}$$

EXERCISES 7

Combine the following radicals:

1. $\sqrt{50}+\sqrt{8}$

2. $\sqrt{48}+\sqrt{27}$

3. $\sqrt{45}-\sqrt{20}$

4. $\sqrt{98}-\sqrt{72}$

5. $\sqrt{75}-\sqrt{12}$

6. $\sqrt{80}+\sqrt{45}$

7. $\sqrt{128}+\sqrt{18}$

8. $\sqrt{108}-\sqrt{48}$

9. $\sqrt{162}-\sqrt{72}$

10. $\sqrt{180}+\sqrt{125}$

11. $2\sqrt{125}-3\sqrt{20}$

12. $5\sqrt{12}-2\sqrt{27}$

13. $6\sqrt{28}+2\sqrt{63}$

14. $2\sqrt{40}-3\sqrt{90}$

15. $9\sqrt{8}-\sqrt{72}$

16. $2\sqrt{98}-5\sqrt{18}$

17. $\sqrt{80}-4\sqrt{20}$

18. $\frac{1}{3}\sqrt{288}+\frac{2}{5}\sqrt{450}$

19. $\sqrt[3]{3}+\sqrt[3]{24}$

20. $\sqrt[3]{54}-\sqrt[3]{16}$

21. $\sqrt[3]{40}+\sqrt[3]{5}$

22. $\sqrt[3]{56}+\sqrt[3]{-7}$

23. $3\sqrt[3]{162}+2\sqrt[3]{48}$

24. $4\sqrt[3]{250}-3\sqrt[3]{432}$

25. $3\sqrt[3]{192}-5\sqrt[3]{24}$

26. $3\sqrt{45}-\sqrt{25}+2\sqrt{5}$

27. $4\sqrt{63}-5\sqrt{28}+3\sqrt{20}$

28. $3\sqrt{98}+\frac{1}{2}\sqrt{48}+\sqrt{50}$

29. $5\sqrt{75}+7\sqrt{108}-6\sqrt{245}$

30. $\sqrt{\frac{1}{2}}+\sqrt{\frac{2}{3}}+\sqrt{\frac{1}{6}}$

31. $8\sqrt{\frac{1}{12}}-\frac{5}{\sqrt{75}}+10\sqrt{\frac{3}{5}}$

32. $3\sqrt{8\frac{1}{3}}-2\sqrt{5\frac{1}{3}}-2\sqrt{1\frac{1}{3}}$

33. $\sqrt{\frac{4}{3}}-\sqrt{\frac{3}{4}}-\sqrt{\frac{1}{12}}$

34. $5\sqrt[3]{2}+4\sqrt[3]{16}+3\sqrt[3]{24}$

35. $2\sqrt[3]{64}-\sqrt[3]{48}-6\sqrt[3]{\frac{1}{36}}$

36. $\sqrt[3]{8}-\sqrt[3]{-40}+\sqrt[3]{125}$

Lesson 10: Multiplication of Multinomials

A multinomial is an expression with multiple terms, such as $a+b-c+d$. In this lesson we will consider multiplying multinomials containing radicals. Note that an expression with two terms is called a **binomial** and an expression with three terms a **trinomial**.

Example 1. Multiply $3\sqrt{2}+2\sqrt{6}-\sqrt{10}$ by $5\sqrt{2}$.

Solution. We will multiply each term of the trinomial by $5\sqrt{2}$, to give:

$$\begin{aligned} 5\sqrt{2}\,(3\sqrt{2}+2\sqrt{6}-\sqrt{10}) &= 15\sqrt{4}+10\sqrt{12}-5\sqrt{20}, \\ &= 30+20\sqrt{3}-10\sqrt{5}. \end{aligned}$$

Example 2. Multiply $\sqrt{45}-\sqrt{32}$ by $\sqrt{20}+\sqrt{18}$.

Solution. Working with radicals is often made easier by simplifying the radicals. For example, we first simplify the expressions in this example:

$$\begin{aligned} \left(\sqrt{45}-\sqrt{32}\right)\left(\sqrt{20}+\sqrt{18}\right) &= \left(3\sqrt{5}-4\sqrt{2}\right)\left(2\sqrt{5}+3\sqrt{2}\right) \\ &= 6\sqrt{25}+9\sqrt{10}-8\sqrt{10}-12\sqrt{4} \\ &= 30+\sqrt{10}-24, \\ &= 6+\sqrt{10}. \end{aligned}$$

Example 3. Evaluate $3x^2-6x-17$ when $x=1+2\sqrt{2}$.

Solution. Since $x=1+2\sqrt{2}$,

$$x^2=(1+2\sqrt{2})^2=1+4\sqrt{2}+8=9+4\sqrt{2}.$$

By substitution:

$$\begin{aligned} 3x^2-6x-17 &= 3(9+4\sqrt{2})-6(1+2\sqrt{2})-17 \\ &= 27+12\sqrt{2}-6-12\sqrt{2}-17, \\ &= 4. \end{aligned}$$

EXERCISES 8

Multiply and simplify :

1. $\sqrt{3}(\sqrt{6}+\sqrt{12})$

2. $\sqrt{5}(\sqrt{5}+\sqrt{15})$

3. $\sqrt{2}(\sqrt{2}+\sqrt{3}-\sqrt{6})$

4. $\sqrt{6}(\sqrt{10}-\sqrt{21}-\sqrt{8})$

5. $\sqrt{6}(3\sqrt{2}-2\sqrt{3}-\sqrt{6})$

6. $\sqrt{10}(2\sqrt{5}-5\sqrt{2}+3\sqrt{15})$

7. $4\sqrt{2}(5\sqrt{2}-6\sqrt{6}-2\sqrt{10})$

8. $3\sqrt{6}(4\sqrt{2}-5\sqrt{3}+3\sqrt{6})$

9. $2\sqrt{12}(3\sqrt{27}-5\sqrt{8}-2\sqrt{15})$

10. $5\sqrt{18}(2\sqrt{12}+3\sqrt{10}+\sqrt{98})$

11. $(\sqrt{3}-\sqrt{2})(\sqrt{3}+\sqrt{2})$

12. $(\sqrt{5}+\sqrt{2})^2$

13. $(3\sqrt{7}+2\sqrt{3})(2\sqrt{7}-5\sqrt{3})$

14. $(5\sqrt{5}+\sqrt{3})(2\sqrt{5}-9\sqrt{3})$

15. $(2\sqrt{2}+3\sqrt{3})^2$

16. $\left(3\sqrt{2}-2\sqrt{2}\right)\left(3\sqrt{2}+2\sqrt{3}\right)$

17. $(\sqrt{12}+2\sqrt{8})(2\sqrt{27}-\sqrt{50})$

18. $(3\sqrt{18}-2\sqrt{75})(2\sqrt{98}-\sqrt{48})$

19. $(3\sqrt{72}-2\sqrt{27})^2$

20. $(4\sqrt{50}+5\sqrt{12})^2$

21. $(\sqrt[3]{16}+\sqrt[3]{9})(\sqrt[3]{4}-\sqrt[3]{3})$

22. $(1+\sqrt[3]{2})(1-\sqrt[3]{2}+\sqrt[3]{4})$

23. $(2\sqrt{3}-3\sqrt{2})^3$

24. $(\sqrt{6}+\sqrt{10}+\sqrt{15})^2$

25. $\sqrt{x}(\sqrt{ax}-\sqrt{bx}+\sqrt{cx})$

26. $(\sqrt{x}+\sqrt{y})(\sqrt{x}-\sqrt{y})$

27. $(\sqrt{a}-\sqrt{b})^2$

28. $\sqrt{ab}(\sqrt{a}+\sqrt{b})$

29. $(\sqrt{n}-\sqrt{1-n})^2$

30. $(\sqrt{x+y}+\sqrt{x-y})^2$

31. $(1+\sqrt{a-3})^3$

32. $\left(x\sqrt{2}-\sqrt{1-2x^2}\right)^2$

Find the value of each of the following:

33. x^2+x+1, when $x=2+\sqrt{2}$

34. x^2-3x-7, when $x=1-\sqrt{3}$

35. $x^2-2x-17$, when $x=1-3\sqrt{2}$

36. x^2+4x+1, when $x=\sqrt{3}-2$

37. $2x^2-x+7$, when $x=\sqrt{3}-\sqrt{2}$

38. $3x^2-2x-5$, when $x=3+2\sqrt{5}$

Lesson 11: Division of Multinomials

Dividing a multinomial involving radicals by a single radical expression, involves dividing term by term. Here are some examples:

Example 1. Divide $4\sqrt{3}-12\sqrt{2}+6\sqrt{6}$ by $2\sqrt{6}$.

Solution. Dividing term by term gives

$$\frac{4\sqrt{3}-12\sqrt{2}+6\sqrt{6}}{2\sqrt{6}} = 2\sqrt{\frac{1}{2}}-6\sqrt{\frac{1}{3}}+3,$$
$$= \sqrt{2}-2\sqrt{3}+3.$$

Rationalizing the Denominator

There is a very clever trick called rationalizing the denominators, that one can use to simplify a ratio of radicals. The objective is to eliminate any radicals in a denominator. This is done by multiplying the top and bottom by the radical we want to remove from the denominator. This process is known as **rationalizing the denominator**. Let's first give a simple example. Simplify the following expression by rationalizing the denominator:

$$\frac{\sqrt{2}}{\sqrt{5}}$$

We multiply top and bottom by $\sqrt{5}$:

$$\frac{\sqrt{2}}{\sqrt{5}} = \frac{\sqrt{2}}{\sqrt{5}} \times \frac{\sqrt{5}}{\sqrt{5}} = \frac{\sqrt{2}\sqrt{5}}{5} = \frac{1}{5}\sqrt{10}.$$

Notice that when we multiply $\sqrt{5}$ by $\sqrt{5}$ we get 5. The purpose of rationalizing the denominator is to remove any radicals from the denominator and move them to the numerator. In example 1 at the top of the page, we would multiply top and bottom by $\sqrt{6}$.

$$\frac{4\sqrt{3}-12\sqrt{2}+6\sqrt{6}}{2\sqrt{6}}=\frac{(4\sqrt{3}-12\sqrt{2}+6\sqrt{6})\cdot\sqrt{6}}{2\sqrt{6}\cdot\sqrt{6}}$$
$$=\frac{12\sqrt{2}-24\sqrt{3}+36}{12}$$
$$=\sqrt{2}-2\sqrt{3}+3.$$

Multiplying Products

Here is an important rule you should be aware of. If we multiply the sum and the difference of two quadratic radicals, we obtain a **rational product**. That means the radicals disappear. For example:

$$\left(\sqrt{a}+\sqrt{b}\right)\left(\sqrt{a}-\sqrt{b}\right)=\left(\sqrt{a}\right)^2-\left(\sqrt{b}\right)^2=a-b,$$

$$\left(4\sqrt{3}+3\sqrt{5}\right)\left(4\sqrt{3}-3\sqrt{5}\right)=\left(4\sqrt{3}\right)^2-\left(3\sqrt{5}\right)^2=48-45=3.$$

In the division of radicals, if the denominator is a binomial quadratic surd of the form $a\sqrt{b}+c\sqrt{d}$, we can rationalize the denominator by multiplying top and bottom by $a\sqrt{b}-c\sqrt{d}$. Note that we multiply by the difference. If the denominator was already a difference, such as $a\sqrt{b}-c\sqrt{d}$ we would multiply top and bottom by the sum: $a\sqrt{b}+c\sqrt{d}$. This trick allows us to get rid of the radicals from the denominator.

Example 2. Simplify:

$$\frac{\sqrt{14}}{2\sqrt{2}-\sqrt{7}}$$

Solution. We will multiply the numerator and the denominator by $2\sqrt{2}+\sqrt{7}$ since the denominator contains $2\sqrt{2}-\sqrt{7}$:

$$\frac{\sqrt{14}}{2\sqrt{2}-\sqrt{7}}=\frac{\sqrt{14}\left(2\sqrt{2}+\sqrt{7}\right)}{(2\sqrt{2}-\sqrt{7})(2\sqrt{2}+\sqrt{7})}=\frac{4\sqrt{7}+7\sqrt{2}}{8-7}=4\sqrt{7}+7\sqrt{2}.$$

EXERCISES 10

Divide and simplify:

1. $\dfrac{3\sqrt{6}-9\sqrt{15}+12\sqrt{21}}{3\sqrt{3}}$

2. $\dfrac{15\sqrt{6}-10\sqrt{10}-25\sqrt{22}}{5\sqrt{2}}$

3. $\dfrac{4\sqrt{2}-6\sqrt{3}+2\sqrt{5}}{\sqrt{2}}$

4. $\dfrac{12\sqrt{5}+18\sqrt{7}+36\sqrt{11}}{2\sqrt{3}}$

5. $\dfrac{5\sqrt{5}+10\sqrt{10}-15\sqrt{15}}{\sqrt{30}}$

6. $\dfrac{7\sqrt{6}-28\sqrt{10}-21\sqrt{26}}{\sqrt{14}}$

7. $\dfrac{3\sqrt{5}+4\sqrt{13}-2\sqrt{19}}{2\sqrt{6}}$

8. $\dfrac{\sqrt{10}-5\sqrt{7}-3\sqrt{15}}{4\sqrt{5}}$

9. $\dfrac{2}{\sqrt{3}-1}$

10. $\dfrac{4}{3-\sqrt{5}}$

11. $\dfrac{3}{\sqrt{3}+\sqrt{2}}$

12. $\dfrac{1+\sqrt{6}}{\sqrt{2}+\sqrt{3}}$

13. $\dfrac{1-\sqrt{10}}{\sqrt{5}+\sqrt{2}}$

14. $\dfrac{3\sqrt{7}+12}{2\sqrt{7}-1}$

15. $\dfrac{\sqrt{5}+2}{\sqrt{5}-2}$

16. $\dfrac{\sqrt{7}-\sqrt{5}}{\sqrt{7}+\sqrt{5}}$

17. $\dfrac{2\sqrt{3}}{3\sqrt{2}-2\sqrt{3}}$

18. $\dfrac{16-\sqrt{35}}{2\sqrt{5}+\sqrt{7}}$

19. $\dfrac{1}{\sqrt{a}+\sqrt{b}}$

20. $\dfrac{\sqrt{x}+\sqrt{y}}{\sqrt{x}-\sqrt{y}}$

Simplify the following expressions:

21. $\dfrac{4\sqrt{3}-3\sqrt{2}}{\sqrt{6}} \div \dfrac{\sqrt{10}}{4\sqrt{3}+3\sqrt{2}}$

22. $\dfrac{\sqrt{2}}{\sqrt{2}+2} \cdot \dfrac{5+4\sqrt{2}}{\sqrt{2}+3}$

23. $\dfrac{\sqrt{2}}{\sqrt{2}-1} + \dfrac{7\sqrt{2}}{2\sqrt{2}-1}$

24. $\dfrac{3\sqrt{2}+2\sqrt{3}}{\sqrt{3}-\sqrt{2}} - \dfrac{5\sqrt{2}-2\sqrt{3}}{\sqrt{3}+\sqrt{2}}$

25. $\dfrac{1}{\sqrt{3}-\sqrt{2}} \cdot \dfrac{4-\sqrt{6}}{1+\sqrt{6}}$

26. $\dfrac{8-2\sqrt{15}}{\sqrt{5}-\sqrt{3}} \div \dfrac{\sqrt{6}}{6+2\sqrt{15}}$

Lesson 12: Radicals of Different Orders

In a previous lesson we learned that $\sqrt[n]{a^m} = a^{\frac{m}{n}}$. The equivalence of a radical and exponent form allows us to change the order of a given radical. For example, in the following we convert a 4th order radical to a square root radical:

$$\sqrt[4]{a^2} = a^{\frac{2}{4}} = a^{\frac{1}{2}} = \sqrt{a}.$$

This was done by dividing top and bottom of the fraction by 2 which allows us to reduce the order by two. Here is another example where we reduce a radical of order 9 to a radical of order 3.

$$\sqrt[9]{a^6} = a^{\frac{6}{9}} = a^{\frac{2}{3}} = \sqrt[3]{a^2}$$

In general, if m, n, and r are any numbers, then

$$\sqrt[nr]{a^{mr}} = a^{\frac{mr}{nr}} = a^{\frac{m}{n}} = \sqrt[n]{a^m}$$

The opposite is also true:

$$\sqrt[n]{a^m} = \sqrt[nr]{a^{mr}}$$

This result can sometimes be used to reduce radicals to a simpler form.

Example 1. Simplify $\sqrt[4]{9}$, $\sqrt[12]{8}$, and $\sqrt[10]{10{,}000}$.

Solution. Writing each of the radicands in exponent form, we have

$$\sqrt[4]{9} = \sqrt[4]{3^2} = \sqrt{3}$$
$$\sqrt[12]{8} = \sqrt[12]{2^3} = \sqrt[4]{2}$$
$$\sqrt[10]{10{,}000} = \sqrt[10]{10^4} = \sqrt[5]{10^2} = \sqrt[5]{100}$$

Note: It is sometimes convenient to think of higher-order radicals as compound radicals; for example:

$$\sqrt[4]{a} = a^{\frac{1}{4}} = \left(a^{\frac{1}{2}}\right)^{\frac{1}{2}} = \sqrt{\sqrt{a}}$$
$$\sqrt[6]{a} = a^{\frac{1}{6}} = \left(a^{\frac{1}{2}}\right)^{\frac{1}{3}} \text{ or } \left(a^{\frac{1}{3}}\right)^{\frac{1}{2}} = \sqrt[3]{\sqrt{a}} \text{ or } \sqrt{\sqrt[3]{a}}$$
$$\sqrt[12]{a} = \sqrt[4]{\sqrt[3]{a}} \text{ or } \sqrt[6]{\sqrt{a}}, \text{ etc.}$$

Example 2. Simplify $\sqrt[6]{27}$ and $\sqrt[8]{25m^2}$.

Solution.

$$\sqrt[6]{27} = \sqrt{\sqrt[3]{27}} = \sqrt{3} \text{ and } \sqrt[8]{25m^2} = \sqrt[4]{\sqrt{25m^2}} = \sqrt[4]{5m}$$

The most important aspect of the above result is that it also allows us to multiply or divide radicals of different orders. To do this we first express the given radicals as radicals of some common order. This process is illustrated in the following example where the two radicals are of second and 3rd order respectively,

Example 3. Multiply $\sqrt{2}$ by $\sqrt[3]{4}$.

Solution. Reducing each of the given radicals to the same lowest order gives

$$\sqrt{2} = \sqrt[6]{2^3} \text{ and } \sqrt[3]{4} = \sqrt[6]{4^2}$$

Since $4 = 2^2$, on multiplying and simplifying, we have

$$\sqrt{2} \cdot \sqrt[3]{4} = \sqrt[6]{2^3} \cdot \sqrt[6]{4^2} = \sqrt[6]{2^3 \cdot 4^2} = \sqrt[6]{2^7} = 2\sqrt[6]{2}$$

EXERCISES 11

Reduce the order of the following radicals:

1. $\sqrt[4]{4}$ **2.** $\sqrt[6]{25}$ **3.** $\sqrt[6]{8}$ **4.** $\sqrt[8]{49}$

5. $\sqrt[6]{81}$ **6.** $\sqrt[9]{64}$ **7.** $\sqrt[8]{16}$ **8.** $\sqrt[12]{125}$

9. $\sqrt[10]{32}$ **10.** $\sqrt[12]{144}$ **11.** $\sqrt[10]{121}$ **12.** $\sqrt[8]{81}$

13. $\sqrt[4]{9x^2y^2}$ **14.** $\sqrt[6]{8a^3}$ **15.** $\sqrt[8]{16m^4n^4}$ **16.** $\sqrt[12]{64a^3b^6c^9}$

Express as radicals of the same lowest order:

17. $\sqrt{3}, \sqrt[3]{5}$ **18.** $\sqrt[4]{7}, \sqrt{9}$ **19.** $\sqrt[4]{2}, \sqrt[5]{3}$ **20.** $\sqrt[3]{4}, \sqrt[4]{3}$

21. $\sqrt[3]{7}, \sqrt[6]{42}$ **22.** $\sqrt[3]{3}, \sqrt[5]{5}$ **23.** $\sqrt{2}, \sqrt[4]{4}, \sqrt[8]{8}$ **24.** $\sqrt[3]{2}, \sqrt[4]{3}, \sqrt[6]{5}$

25. $\sqrt{3}, \sqrt[3]{5}, \sqrt[4]{7}$ **26.** $\sqrt{2a}, \sqrt[3]{3b^2}$ **27.** $\sqrt[4]{x^3}, \sqrt[6]{3x^5}$ **28.** $\sqrt[3]{3m}, \sqrt[4]{2n^3}$

Multiply the following expressions:

29. $\sqrt{5} \cdot \sqrt[3]{3}$ **30.** $\sqrt[3]{2} \cdot \sqrt[4]{3}$ **31.** $\sqrt{7} \cdot \sqrt[4]{11}$

32. $\sqrt[3]{2} \cdot \sqrt[6]{6}$ **33.** $\sqrt[4]{4} \cdot \sqrt[6]{5}$ **34.** $\sqrt{2} \cdot \sqrt[10]{7}$

35. $\sqrt{ab} \cdot \sqrt[5]{a^4 b}$ **36.** $\sqrt[3]{x^2} \cdot \sqrt[7]{x^5}$ **37.** $\sqrt[4]{2m^3 n} \cdot \sqrt[12]{5m^5 n^9}$

38. $\sqrt{3a} \cdot \sqrt[3]{9a^2}$ **39.** $\sqrt[5]{a^4 b^3} \cdot \sqrt[6]{a^3 b^5}$ **40.** $\sqrt[5]{3x^4 y^2} \cdot \sqrt[10]{7x^2 y^6}$

Lesson 13: Imaginary Numbers

One problem that new students will find quite surprising is what to do with the square root of a negative number. At first glance it may not seem to be a problem, but one soon realizes that there doesn't appear to be an answer. For example, what would we expect the square root of -4 to be? If can't be $+2$ since $2 \times 2 = 4$ and it can't be -2 since $-2 \times -2 = 4$.

Since the square roots of negative quantities occur quite often in mathematics and particulary in engineering, the solution to the problem was to invent a new kind of number called an **imaginary number**.

Definition. *The square root of a negative number is called an* ***imaginary number.***

Examples: $\sqrt{-4}, \sqrt{-3}, \sqrt{-9a^2}$ are imaginary numbers.

To help distinguish imaginary numbers, the ordinary numbers which we deal with everyday are referred to as **real numbers**.

By the law of radical multiplication (lesson 6), it is always possible to write any imaginary number, such as $\sqrt{-a}$, in the form $\sqrt{-1 \times a} = \sqrt{a}\sqrt{-1}$. As a result, we will give the following definition.

Definition. The imaginary number $\sqrt{-1}$ is the unit imaginary value and is represented by the letter i.

Any imaginary number can be expressed in terms of i. Thus, for the examples given above, we have

$$\sqrt{-4} = \sqrt{4}\sqrt{-1} = 2i, \quad \sqrt{-\frac{3}{4}} = \sqrt{\frac{3}{4}}\sqrt{-1} = \frac{\sqrt{3}}{2}i,$$
$$\sqrt{-9a^2} = \sqrt{9}a^2\sqrt{-1} = 3ai.$$

From the above definition of i, it follows that $i^2 = -1, i^3 = i^2 \cdot i = -i, i^4 = i^2 \cdot i^2 = (-1)(-1) = +1$, etc. From this we see a pattern such that higher powers of the imaginary unit is equal to one of the four values $i, -1, -i, +1$ according to the following table:

$$i = i^5 = i^9 \quad = \cdots \text{ etc. } = i,$$
$$i^2 = i^6 = i^{10} = \cdots \text{ etc. } = -1,$$
$$i^3 = i^7 = i^{11} = \cdots \text{ etc. } = -i,$$
$$i^4 = i^4 = i^{12} = \cdots \text{ etc. } = +1.$$

EXERCISES 11

Express each of the following in terms of i and simplify:

1. $\sqrt{-4}$ **2.** $\sqrt{-25}$ **3.** $\sqrt{-64}$ **4.** $\sqrt{-8}$
5. $\sqrt{-12}$ **6.** $2\sqrt{-9}$ **7.** $3\sqrt{-49}$ **8.** $2\sqrt{-50}$
9. $5\sqrt{-18}$ **10.** $4\sqrt{-45}$ **11.** $\sqrt{-4a^2}$ **12.** $\sqrt{-32a^2x^2}$
13. $\sqrt{-20x^4}$ **14.** $\sqrt{-9x^3}$ **15.** $\sqrt{-8a^5}$

Given that $i = \sqrt{-1}$, express each of the following in terms of i or a real number:

16. i^7 **17.** i^{12} **18.** i^{17} **19.** i^{25} **20.** i^{13}
21. i^{10} **22.** i^{19} **23.** i^{82} **24.** i^{57} **25.** i^{100}
26. $(-i)^3$ **27.** $(-i^3)^4$ **28.** $(-i)^6$ **29.** $(-i^7)^7$ **30.** $(-i^4)^3$

Lesson 14: Equations Involving Radicals

An equation containing the unknown value such as x inside a radical is called a radical equation.

Examples: $\sqrt{x}=7$, $2\sqrt[3]{x}=5$, and $\sqrt{x^2-5}=x+1$ are all radical equations, whereas $x\sqrt{2}=3$ is not since the x is outside the radical.

The objective is to solve such equations for the unknown quantity, x. The trick is to square (or raise to what ever convenient power it needs to be) both sides as in the following example:

Example 1. Solve $\sqrt{x-2}=7$.

Solution. Squaring both sides of this equation, we have

$$\begin{aligned}\sqrt{x-2}\sqrt{x-2} &= 7\cdot 7\\ x-2 &= 49\\ x &= 51.\end{aligned}$$

Note: Not all radical equations have solutions. For example, consider $\sqrt{x}=-5$. By squaring both sides, it appears that $x=25$ is a solution. But if we substitute into x the equation doesn't seem to work because $\sqrt{25}$ equals 5 and not -5

Such apparent solutions are called *extraneous* solutions and should be rejected when solving radical equations. Hence, it is *necessary* to check all solutions of radical equations.

Example 2. Solve $\sqrt{4x-11}=2\sqrt{x}-1$.

Solution. Squaring both sides of this equation gives

$$\begin{aligned}4x-11 &= 4x-4\sqrt{x}+1,\\ 4\sqrt{x} &= 12.\\ \sqrt{x} &= 3.\end{aligned}$$

Squaring this new equation, we have

$$x=9.$$

Hence, $x = 9$ is the only *possible* solution. Check.

$$\sqrt{4\cdot 9-11} = 2\sqrt{9}-1,$$
$$\sqrt{25} = 2\cdot 3-1,$$
$$5 = 5.$$

Thus, $x = 9$ is a valid solution.

Example 3. Solve $2\sqrt{x+6}+3\sqrt{x+1}=0$.

Solution. Transposing and squaring, we have

$$2\sqrt{x+6} = -3\sqrt{x+1}$$
$$4(x+6) = 9(x+1)$$
$$5x = 15$$
$$x = 3$$

Hence, $x = 3$ is the only *possible* solution.

Check.

$$2\sqrt{3+6}+3\sqrt{3+1} = 0$$
$$2\cdot 3+3\cdot 2 = 0$$
$$12 = 0$$

Which makes no sense. Thus, $x = 3$ is extraneous and there are *no* solutions to the given radical equation.

EXERCISES 12

Solve:

1. $\sqrt{x-4}=3$ **2.** $\sqrt{2x-1}=5$ **3.** $4+\sqrt{x+3}=0$

4. $\sqrt{x^2-1}=x-1$ **5.** $\sqrt{9x^2-5}=3x-5$ **6.** $\sqrt{5x+6}=-2$

7. $\sqrt{4x^2+9}=2x+1$ **8.** $\sqrt{4x^2+17}=2x+1$ **9.** $\sqrt[3]{x+2}=-3$

10. $\sqrt[3]{2x-1}=4$ **11.** $x+\sqrt{x^2+1}=1$ **12.** $3x+\sqrt{9x^2-8}=2$

13. $3x+\sqrt{9x^2-5}=1$ **14.** $x-\sqrt{x^2-9}=1$ **15.** $\sqrt{5x-1}=2\sqrt{x+1}$

16. $\sqrt[3]{9x-4}=2\sqrt[3]{x-1}$ **17.** $\sqrt{x}+\sqrt{x+5}=5$ **18.** $\sqrt{x-4}+3=\sqrt{x+11}$

19. $\sqrt{2x-1}+2=\sqrt{2x+11}$ **20.** $\sqrt{x-5}+\sqrt{x+10}=3$ **21.** $\sqrt{x+4}+\sqrt{x+16}=2$

22. $\sqrt{5x}+\sqrt{5x-9}=1$ **23.** $\sqrt{x+31}-\sqrt{x+14}=1$ **24.** $\sqrt{x+12}-\sqrt{x-12}=2$

Lesson 15: Computing Square Roots

The following two lessons are more for historical value as it discusses how to compute square roots manually without the use of an electronic calculator. Lesson 17 will go over how electronic calculators compute square roots. The technique is of interest however.

Consider the following table of squares:

$1^2 = 1$ $9^2 = 81$	$0.1^2 = 0.01$ $0.9^2 = 0.81$
$10^2 = 1'00$ $99^2 = 98'01$	$0.01^2 = 0.00'01$ $0.99^2 = 0.98'01$
$100^2 = 1'00'00$ $999^2 = 99'80'01$	$0.001^2 = 0.00'00'01$ $0.999^2 = 0.99'80'01$

Looking at this table suggests the following rule:

> Rule for Evaluation of Quadratic Radicals
>
> If a number representing a square is separated into groups of two figures each, beginning at the decimal point, the square root of the number will have as many figures as there are groups.

Thus, the square root of $729\,(=7'29)$ will be a two-figured number.

Let x be the tens' digit and y the units' digit of the number 729. It follows that

$$729 = (10x + y)^2,$$

or

$$729 = 100x^2 + 20xy + y^2.$$

Since x and y are positive digits, it must be true that $100x^2$ cannot be greater than 700. Also, since x must be an integer, we see that

$x^2 = 4$, or $x = 2$. Hence, $400 = 100x^2$, and if we subtract from the above equation,

$$329 = 20xy + y^2$$

However, $x = 2$, so this equation becomes

$$329 = 40y + y^2,$$

$$\text{or} \qquad 329 = (40 + y)y.$$

Since y represents a digit, 40 may be taken as a trial number for $(40 + y)$. Hence dividing 329 by 40 we see that y might be 8. However, since $(40 + 8)8 = 384$, this value of y is too large; so we try a smaller value, $y = 7$. We now have $(40 + 7)7 = 329$ which does satisfy the above equation. Hence, since $x = 2$ and $y = 7$, the square root of 729 is 27.

Before the advent of electronic calculators, students were taught a manual method for computing square roots based on the above analysis. The method uses what looks like long division but it isn't long division as we know it. We'll refer to it as a table. An example is shown below. I have split the process into four steps. The first thing to do is to group the number we want to find the square root of into groups of two digits per group starting from the right. With 729, there are two groups, with the left most group containing a single digit. This grouping is shown as $\overline{7}\ \overline{29}$. By the rule we gave above, since there are two groups, the square root will contain two digits.

Step One: In the first step we look at the first number group on the left, in this case it is the single digit 7. We next determine the largest number whose perfect square is less than or equal to 7. In this case the largest perfect square less than 7 is 4, and obtain the corresponding square root which will be 2. We will need this number. Note that the next perfect square after 4 is 9 which is greater than 7, hence we must pick 4. We can now start constructing our table, see the table on the next page.

We write down the 2 (the square root of 4 from the last paragraph) in two places, above the 7 and to the left of the 7. We place the square of 2, which is 4, beneath the 7 and subtract the square from the 7 giving us 3. We draw a line underneath the 4 and write the 3 underneath the 4 (this bit is similar to long division). We copy the number 2 to the left of the 7 and place it beneath the 2. This completes step one.

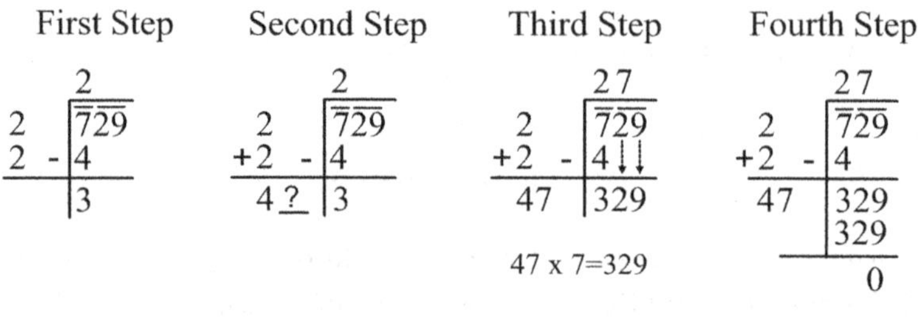

The second and third steps are repeated multiple times until the remainder is 0 or we have accumulated sufficient digits to approximate the square root if it is a decimal.

Second Step: In the second step we add the left most two digits which is effectively doubling the 2 to give us a value of 4. Place the 4 underneath the 2. It is recommended that this column be moved to the left to reserve room for a second digit which we will compute shortly.

Third Step: In the third step we bring down the next group of numbers, in this case 29, to be next to the reminder 3, giving us a new number 329. This number will, by the design of the table, be adjacent to the 4 and the space we reserved for a second digit. Let us designate the digit in the blank space with a question mark '?'. This is shown in the table as '4 ?'. The digit in the blank space, ?, will be the next digit in the solution.

The digit in the blank space is computed such that the product 4? × ? is equal to or less than the value 329. The number that satisfies this will be 47 since $47 \times 7 = 329$ which is exactly the number we seek. We multiply 47×7 and place this value underneath the 329. That completes the third step. We now compute the reminder by subtracting one from the other, but since they are the same the reminder is zero.

Fourth Step: Because the remainder is zero this completes the computation. If the reminder were not zero we would repeat the second and third steps again based on the remainder.

The square root can be read from the top set of digits, in this case 27 which is correct. Let's look at some more examples.

Example 1. Find the square root of 424 to 3 decimal places.

Solution.

```
              20.591
            ___________
   2       |424.000000
  +2     - |4
  ------------
  40       | 24
  + 0    - | 00
  ------------
  405      |  2400
    5    - |  2025
  ------------
  4109     |   37500
     9 -   |   36981
  ------------
  41181    |     51900
      1    |     41181
  -----------------
                etc.
```

Example 2. Find the square root of 10.416 correct to the nearest hundredth.

Solution.

```
             3.227
            _________
   3       |10.416000
  +3     - | 9
  ------------
  62       | 141
  + 2    - | 124
  ------------
  642      |  1760
    2    - |  1284
  ------------
  6447     |   47600
         - |   45129
  -----------------
                2471
```

When the number is not a perfect square, this process of computation may be continued to any desired accuracy. In this case we stop once we have three decimal places. Hence, the square root of 10.116 correct to the nearest hundredth is 3.23.

At first glance this seems like a complicated method but with practice it can be done quite quickly. Although perhaps only of historical interest today given the prevalence of electronic calculators, there are many YouTube videos that describe this algorithm. If you'd like to try out the method you can try doing the following exercises.

EXERCISES 13

Find the square roots of the following numbers:

1. 8.8804	**2.** 790321	**3.** 5012.64	**4.** 299.29
5. 0.881721	**6.** 76.0384	**7.** 11881	**8.** 4.9729
9. 0.217009	**10.** 0.029241	**11.** 0.000484	**12.** 3672.36

Find the square root of the following numbers correct to the nearest hundredth:

13. 17.06	**14.** 1.887	**15.** 23.95	**16.** 9.4061
17. 0.801	**18.** 67.85	**19.** 0.0515	**20.** 5.001
21. 20.0102	**22.** 12.3456	**23.** 16.6666	**24.** 253

Lesson 16: Using Tables

Before the advent of electronic calculators, math and engineering students often carried with them booklets containing tables of numbers. One such table in the booklet was a table of square roots. The table on page 43 is such an example and lists the squares and square roots of the numbers from 1 to 100. For example, to find the square of 29, locate the number 29 in the column headed 'Number' and then read off the square in the column immediately to the right. The value of 29^2, namely 841, appears to the right in the column headed 'Square.'

What is most important to realize is that any number in the column headed 'Square' has its square root in the column headed 'Number.' Thus, the square root of 6724 is 82. Here is another example, the square root of 29 is given in the second column to the right opposite 29 and is found to be 5.385, correct to the nearest thousandth.

Since the table doesn't contain every possible square root, it is often necessary to modify or simplify the radicand in order that it may be found in the table. The following examples indicate some of the tricks employed.

Example 1. Evaluate $\sqrt{140}$, using tables.

Solution: Although the number 140 is not in the table, we can first simplify the radical. Thus,

$$\sqrt{140} = \sqrt{4 \cdot 35} = 2\sqrt{35} = 2\,(5.916) = 11.832.$$

Since 35 is in the table we can compute its square root and multiply by 2 to get the final answer.

Example 2. Evaluate $\sqrt{2\frac{1}{2}}$, using tables.

Solution: We have $\sqrt{2\frac{1}{2}} = \sqrt{\frac{5}{2}}$ and although both $\sqrt{5}$ and $\sqrt{2}$ are in the table, it is recommended we simplify further. Thus,

$$\sqrt{2\frac{1}{2}} = \sqrt{\frac{5}{2}} = \sqrt{\frac{10}{4}} = \frac{1}{2}\sqrt{10} = \frac{1}{2}(3.162) = 1.581.$$

Instead of having to get the square root of two numbers, $\sqrt{5}$ and $\sqrt{2}$, we now only need to lookup the square root of a single number $\sqrt{10}$.

Example 3. Evaluate $\sqrt{1.71}$, using tables.

Solution: First removing decimals, we have

$$\sqrt{1.71} = \sqrt{\frac{171}{100}} = \frac{1}{10}\sqrt{171} = \frac{1}{10}\sqrt{9 \cdot 19} = \frac{3}{10}\sqrt{19} = 0.3\,(4.359) = 1.3077.$$

Example 4. Evaluate $\frac{1}{\sqrt{2}+1}$, using tables.

Solution: Although the $\sqrt{2}$ can be found in the tables, it is recommended to first simplify. Thus,

$$\frac{2}{\sqrt{2}+1} = \frac{2(\sqrt{2}-1)}{(\sqrt{2}+1)(\sqrt{2}-1)} = \frac{2(\sqrt{2}-1)}{2-1} = 2\,(1.414-1) = 0.828.$$

EXERCISES 14

Find the square roots of the following numbers in the table:

1. 19 **2.** 94 **3.** 66 **4.** 41 **5.** 87

6. 529 **7.** 9216 **8.** 3364 **9.** 1936 **10.** 5776

Using the table, find the square roots of the following numbers:

11. 228 **12.** 369 **13.** 175 **14.** 147

15. 605 **16.** $3\frac{2}{5}$ **17.** $2\frac{3}{4}$ **18.** $20\frac{4}{7}$

19. $34\frac{2}{3}$ **20.** $14\frac{2}{5}$ **21.** 1.8 **22.** 8.91

23. 3.25 **24.** 2.94 **25.** 0.608

Simplify and evaluate, using the table:

26. $\frac{\sqrt{5}}{\sqrt{3}-\sqrt{2}}$ **27.** $\frac{3+\sqrt{5}}{3-\sqrt{5}}$ **28.** $\frac{7}{2\sqrt{2}+1}$ **29.** $\frac{5(2+\sqrt{6})}{\sqrt{2}+2\sqrt{3}}$

30. $\frac{1+\sqrt{\frac{1}{2}}}{1+\sqrt{\frac{1}{3}}}$

TABLE OF SQUARES AND SQUARE ROOTS

Number	Square	Square Root	Number	Square	Square Root
1	1	1.000	51	2601	7.141
2	4	1.414	52	2704	7.211
3	9	1.732	53	2809	7.280
4	16	2.000	54	2916	7.348
5	25	2.236	55	3025	7.416
6	36	2.450	56	3136	7.483
7	49	2.646	57	3249	7.550
8	64	2.828	58	3364	7.616
9	81	3.000	59	3481	7.681
10	100	3.162	60	3600	7.746
11	121	3.317	61	3721	7.810
12	144	3.464	62	3844	7.874
13	169	3.606	63	3969	7.937
14	196	3.742	64	4096	8.000
15	225	3.873	65	4225	8.062
16	256	4.000	66	4356	8.124
17	289	4.123	67	4489	8.185
18	324	4.243	68	4624	8.246
19	361	4.359	69	4761	8.307
20	400	4.472	70	4900	8.367
21	441	4.583	71	5041	8.426
22	484	4.690	72	5184	8.485
23	529	4.796	73	5329	8.544
24	576	4.899	74	5476	8.602
25	625	5.000	75	5625	8.660
26	676	5.099	76	5776	8.718
27	729	5.196	77	5929	8.775
28	784	5.292	78	6084	8.832
29	841	5.385	79	6241	8.888
30	900	5.477	80	6400	8.944
31	961	5.568	81	6561	9.000
32	1024	5.657	82	6724	9.055
33	1089	5.745	83	6889	9.110
34	1156	5.831	84	7056	9.165
35	1225	5.916	85	7225	9.220
36	1296	6.000	86	7396	9.274
37	1369	6.083	87	7569	9.327
38	1444	6.164	88	7744	9.381
39	1521	6.245	89	7921	9.434
40	1600	6.325	90	8100	9.487
41	1681	6.403	91	8281	9.539
42	1764	6.481	92	8464	9.592
43	1849	6.557	93	8649	9.644
44	1936	6.633	94	8836	9.695
45	2025	6.708	95	9025	9.747
46	2116	6.782	96	9216	9.798
47	2209	6.856	97	9409	9.849
48	2304	6.928	98	9604	9.899
49	2401	7.000	99	9801	9.950
50	2500	7.071	100	10000	10.000

Lesson 17: How Calculators do Square Roots

There are quite a few ways in which square roots can be calculated. Some methods are very old and go all the way back to the Babylonians. Heron, who lived in Alexandria probably between 10 and 85 AD, wrote a book called Metrica that describes among other things a method to compute the square root of any positive number. Heron described the method in words but in modern notation the formula he used was:

$$a_1 = \frac{1}{2}\left(a + \frac{A}{a}\right)$$

where A is the number we want to find the square root of and a is a guess for what the square root might be. The value on the left, a_1 is an estimate for the square root. However the formula can be used a second time where instead of using a as our guess we now use a_1. The second calculation is shown below:

$$a_2 = \frac{1}{2}\left(a_1 + \frac{A}{a_1}\right)$$

a_2 is an even better estimate for the square root of A. This can be repeated as many times as you like to make the estimate more and more accurate. In practice it only takes a few iterations to get a good estimate, especially if the initial guess is reasonably close to begin with.

Heron's method is probably from an even older Babylonian method. Let's try an example by looking at an easy problem, say computing the square root of 25. Let's start with a guess of 10, which is clearly a bad guess since 10 times 10 is 100. In practice you'd pick a guess whose square is just above the value we're trying to get the square root of. In this case a better guess would be 6, since $6 \cdot 6$ is just above 25. But let's try the 10 nevertheless. We will insert 10 into the first formula:

$$a_1 = \frac{1}{2}\left(10 + \frac{25}{10}\right) = 6.25$$

We'll now insert the answer, 6.25 back into the same formula:

$$a_2 = \frac{1}{2}\left(6.25 + \frac{25}{6.25}\right) = 5.125$$

Let's do it once more:

$$a_3 = \frac{1}{2}\left(5.125 + \frac{25}{5.125}\right) = 5.001524$$

After three 'iterations' we have get quite close to the actual answer of 5. If we did the calculation once more we get even closer at 5.00000023.

The method works quite well, but why does it work? Heron doesn't seem to indicate in his book how the method actually worked. The Arabs appear to have been familiar with the technique (Sharaf al-Din al-Tusi, 1250 AD) but it was later mathematicians such as Newton and Raphson who formalized it and ultimately gave us a reason why the trick worked. Today the method is known as the Newton-Raphson's method and it works like this.

Let's say we have a function like $g(x) = a$, for example $x^2 = a$. We can rearrange this equation to be $g(x) - a = 0$. In general we have a function that is of the form $f(x,a) = 0$. We can plot the left-hand side as a function of x so that when the function, $f(x,a)$ passes through the x axis we're at the point where $f(x,a) = 0$. An example will make this clearer.

Let's consider the function $y = x^2 - 25$ and plot y versus x as shown below. I'm only plotting the graph using positive values for x.

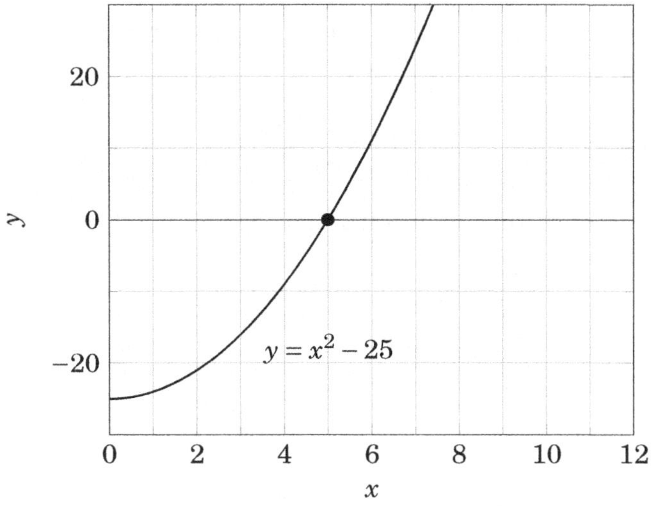

You can see the line going through the $y = 0$ axis marked with a dot. This is when $x^2 - 25 = 0$. This is only true when $x = 5$ because $5 \times 5 - 25 = 0$. If we could find the location of that point marked by a black dot that would effectively be the square root. This is what the Newton-Raphson method (or Heron's method) actually does.

In the following description you should also look at the plot on the next page in order to follow the argument.

The method works by starting with a guess, marked x_1, and at the guess we find the slope of the curve at the point marked A. We extend the slope all the way to the x axis, marked x_2. This is our new estimate.

From this point we start again, find the slope, extend the slope, etc. You can probably see that each time we perform this procedure we get closer and closer to the point marked by the black dot which is our solution. Given this geometrical view we can turn the procedure into a formula. The slope for example is given by:

$$\frac{df}{dx_1} = \frac{f(x_1)}{x_1 - x_2}$$

That is the vertical distance divided by the horizontal distance (rise over run).

If we designate the first guess with the subscript x_k and the new estimate for x as x_{k+1}, we can rewrite the slope as:

$$\frac{df}{dx_k} = \frac{f(x_k)}{x_k - x_{k+1}}$$

We can rearrange this equation to get x_{k+1} on one side to give:

$$x_{k+1} = x_k - \frac{f(x_k)}{df/dx_k}$$

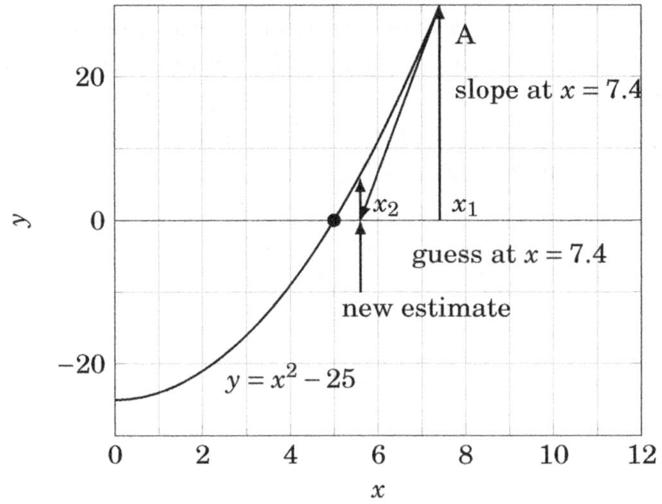

Let's apply this equation to the square root problem. We start with $x^2 = a$, that is we want to find the square root of a. We rearrange it to $x^2 - a = 0$. The slope for this equation is given by differentiating with respect to x which gives:

$$\frac{df}{dx} = 2x$$

We also note that $f(x_k)$ is just $x^2 - a$. Substituting these into the equation above gives:

$$x_{k+1} = x_k - \frac{x^2 - a}{2x_k}$$

Rearranging this gives:

$$x_{k+1} = \frac{1}{2}\left(x_k + \frac{a}{x_k}\right)$$

This is, of course, Heron's formula. How the ancients came up with this method we don't really know. Usually the Greeks would give detailed proofs for all their results but in this case they didn't. It is possible that Archimedes knew of this trick since he computed the cube root of 3, and perhaps he had a proof which is now lost although there is still a lot of debate on this matter.

Note that the formula involves a division. For an electronic calculator the division operation can be slow. If we could just use addition and multiplication the calculations could be done faster. One clever trick that can be used is instead of trying to find the square root of $\sqrt{a}$, we instead try to find the value of $1/\sqrt{a}$. Let:

$$x = \frac{1}{\sqrt{a}}$$

So that:

$$\frac{1}{x} = \sqrt{a}$$

Square both sides:

$$\frac{1}{x^2} = a$$

Our function, $f(x)$ then becomes:

$$f(x) = \frac{1}{x^2} - a$$

Now that we have it in the right form, let's derive the Newton-Raphson equation. First we'll compute the derivative df/dx by differentiating $f(x)$ with respect to x:

$$\frac{df}{dx} = -\frac{2}{x^3}$$

Next we'll apply the Newton-Raphson equation:

$$x_{k+1} = x_k - \frac{f(x_k)}{df/dx_k}$$

Inserting $f(x)$ and df/dx into this equation gives us:

$$x_{k+1} = x_n - \frac{\frac{1}{x_k^2} - a}{-2\frac{1}{x_k^3}}$$

Simplifying this leads to our result:

$$= x_k - \frac{x_k - x_k^3 a}{-2} = \frac{1}{2} x_k \left(3 - x_k^2 a\right)$$

There is still one division by 2 but modern computers have special ways of dividing by 2 very fast. This algorithm let's us compute $1/\sqrt{a}$, but we want $\sqrt{a}$. If you remember from Lesson 11 we talked about rationalizing a radical. This was done to remove a radical from the denominator and involved multiplying top and bottom by the radical. If we do that to $1/\sqrt{a}$ we get:

$$\frac{1}{\sqrt{a}} = \frac{1}{\sqrt{a}}\frac{\sqrt{a}}{\sqrt{a}} = \frac{\sqrt{a}}{a}$$

Hence we can recover $\sqrt{a}$ by multiplying $1/\sqrt{a}$ by a. The algorithm is is therefore apply the iteration a few times then multiply the result by a.

If may seem that this alterative approach will be slower because there are four multiplications in total (including the last multiplication by a) compared to just one division in the original method. However, the speed difference between multiplication and division, depending on the particular computer, can be quite significant, though with more modern computer chips the difference is less pronounced. The web site at cppbenchmarks[1] shows a comparison of multiplication and division and on an Intel i5, which is quite a common computer chip, division is just over 5 times slower than multiplication. Hence we will gain a modest 25% speed increase by using the reciprocal of the square root. Electronic calculators use calculator chips that are likely to be much worse at division and in such cases the reciprocal algorithms will be even faster.

Although not all electronic calculators will compute square roots this way, a lot of them will.

[1] https://cppbenchmarks.wordpress.com/2020/11/10/float-division-vs-multiplication-speed/

Lesson 18: Applications of Square Root

Let's now look at some application of the square root.

One of the most famous discoveries in geometry was first proved by the Greek mathematician Pythagoras in around 550 BC. It is as follows:

> Pythagoras' Theorem
>
> In any right triangle the square on the hypotenuse is equal to the sum of the squares on the other two sides.

If a and b represent the length of the sides of a right triangle and c represents the length of the hypotenuse, this relation states that

$$c^2 = a^2 + b^2$$

From this equation it is also evident that

$$a^2 = c^2 - b^2, \text{ or } b^2 = c^2 - a^2.$$

Example 1. Find the hypotenuse of a right triangle whose sides are 24 meters and 45 meters respectively.

Solution. Let c represent the length of the hypotenuse; then using Pythagoreans' theorem, we get

$$c^2 = 21^2 + 45^2 = 576 + 2025 = 2601$$

Extracting the square root, we find $c = \sqrt{2601} = 51$. Hence, the hypotenuse is 51 meters long.

Example 2. A ladder 30 feet long just reaches the lower edge of a window when the foot of the ladder is 9 feet from the wall. Find, to the nearest tenth of a foot, how high the window is above the ground.

Solution. From the Pythagoreans' theorem,

$$a = \sqrt{30^2 - 9^2} = \sqrt{(30+9)(30-9)} = \sqrt{39 \cdot 21} = \sqrt{3 \cdot 13 \cdot 3 \cdot 7} = 3\sqrt{91}$$

From the table of square roots,

$$a = 3(9.539) = 28.617.$$

Hence, the window is 28.6 feet above the ground.

EXERCISES 15

In the following exercises a, b, and c represent the sides of a right triangle; c is the hypotenuse. You can use a calculator to help you. Find the missing side:

1. $a = 6, b = 8$ **2.** $a = 30, c = 34$ **3.** $b = 10, c = 26$

4. $a = 20, b = 21$ **5.** $b = 36, c = 45$ **6.** $a = 40, c = 85$

7. Find, to the nearest tenth of an inch, the diagonal of a square whose area is 389 square inches.

8. Find, to the nearest tenth of a cm, the diameter of a circle whose area is 99 square cm. (Assume $\pi = \frac{22}{7}$.)

9. A rope attached to the top of a flagpole is 112 feet long. When the rope is pulled tight, the loose end reaches the ground at a point 15 feet from the foot of the pole. How high is the flagpole?

10. Two vertical poles 20 meters and 30 meters high respectively are 60 meters apart. A rope is to be stretched from the top of one pole to a point on the ground halfway between the poles, and then to the top of the other pole. How long a rope is needed?

11. A telephone pole 32 feet high stands 100 feet away from a building 65 feet high. How long a wire is needed to stretch from the top of the pole to the top of the building?

12. If $m^2 + n^2$ represents the hypotenuse of a right triangle and $m^2 - n^2$ represents one of the sides, what expression in terms of m and n represents the third side of the triangle?

Review Exercises

A

Simplify the following radical expressions:

1. $\sqrt{242}$ **2.** $\sqrt{108}$ **3.** $\sqrt{12a^3}$

4. $6\sqrt{200}$ **5.** $\sqrt{98(x-y)^2}$ **6.** $3m^2n\sqrt{180m^5n^3}$

7. $\sqrt{(a^2-b^2)(a+b)}$ **8.** $5\sqrt[3]{-96}$ **9.** $\sqrt[3]{64a^5b^3c^2}$

10. $\sqrt{\frac{7a^3}{18d^5}}$ **11.** $6\sqrt{5\frac{1}{3}}$ **12.** $\sqrt{\frac{4ax}{27by}}$

13. $\frac{12}{7\sqrt{3}}$ **14.** $\sqrt[3]{13\frac{1}{2}}$ **15.** $\sqrt[3]{\frac{8a^4}{3x^2}}$

Combine:

16. $3\sqrt{50}-2\sqrt{18}$ **17.** $x\sqrt{8}+4\sqrt{2x^2}$

18. $\sqrt{150}-6\sqrt{2\frac{2}{3}}$ **19.** $2\sqrt[3]{16}+\sqrt[3]{-54}$

20. $\sqrt[3]{\frac{1}{2}}-\sqrt[3]{\frac{4}{27}}$ **21.** $\sqrt{2a^2}-\sqrt{2b^2}$

22. $\sqrt{3\frac{1}{5}}+\sqrt{1\frac{1}{2}}-\sqrt{11\frac{1}{4}}$ **23.** $3\sqrt{245}-3\sqrt{338}-2\sqrt{125}$

24. $\sqrt{72}+\frac{\sqrt{243}}{3}-\frac{8}{\sqrt{2}}+3\sqrt{\frac{1}{3}}$ **25.** $\sqrt[3]{1\frac{1}{3}}+\sqrt[3]{16}+\sqrt[3]{4\frac{1}{2}}$

26. $\sqrt{a^2b}-\sqrt{b^3}+\sqrt{bc^2}$ **27.** $\sqrt{a^3-a^2b}-\sqrt{ab^2-b^3}$

28. $\sqrt{xy^2}+\sqrt{x^3y^4}+\sqrt{x^5y^6}$ **29.** $\sqrt[3]{p^7}+\sqrt[3]{8p^4q^3}+\sqrt[3]{pq^6}$

30. $\sqrt{\frac{yz}{x}}+\sqrt{\frac{zx}{y}}+\sqrt{\frac{xy}{z}}$ **31.** $\sqrt{\frac{a}{b}+\frac{b}{a}+2}-\sqrt{\frac{b}{a}}$

32. $\sqrt{ax^2+2ax+a}+\sqrt{ax^2-2ax+a}$ **33.** $\sqrt{(x+2)(x^2-4)}-\sqrt{x^3-2x^2}$

B

Multiply:

1. $4\sqrt{\frac{3}{2}}\cdot 3\sqrt{\frac{3}{8}}$ **2.** $3\sqrt{27}\cdot 2\sqrt{75}$

3. $4\sqrt[3]{12}\cdot 5\sqrt[3]{18}$

4. $\frac{2}{5}\sqrt[3]{\frac{9}{16}}\cdot\frac{2}{9}\sqrt[3]{\frac{3}{4}}$

5. $2\sqrt{27}\cdot 3\sqrt{21}\cdot 5\sqrt{28}$

6. $\frac{1}{2}\sqrt{\frac{4}{5}}\cdot\frac{2}{3}\sqrt{\frac{5}{8}}\cdot\frac{3}{5}\sqrt{\frac{1}{3}}$

7. $2\sqrt{3}(2\sqrt{15}-3\sqrt{12}+4\sqrt{21})$

8. $\sqrt{abc}(\sqrt{ab}+\sqrt{bc}+\sqrt{ac})$

9. $(\sqrt{2}+\sqrt{5})(\sqrt{2}+3\sqrt{5})$

10. $(3\sqrt{5}-2\sqrt{10})^2$

11. $(\sqrt[3]{2}+\sqrt[3]{4})(\sqrt[3]{2}-\sqrt[3]{4})$

12. $(\sqrt{7}-1)^2+2(\sqrt{7}-1)+5$

Divide:

13. $4\sqrt{51}\div\sqrt{34}$

14. $5\sqrt{228}\div 2\sqrt{19}$

15. $18\sqrt[3]{13}\div 3\sqrt[3]{26}$

16. $10\sqrt[3]{25}\div 5\sqrt[3]{40}$

17. $(8\sqrt{35}-2\sqrt{15}-4\sqrt{20})\div 2\sqrt{5}$

18. $(4\sqrt{18}+3\sqrt{60}-9\sqrt{2})\div 3\sqrt{12}$

19. $\dfrac{6}{\sqrt{7}-2}$

20. $\dfrac{5\sqrt{5}}{4\sqrt{2}-3\sqrt{3}}$

21. $\dfrac{\sqrt{27}-\sqrt{18}}{\sqrt{75}-\sqrt{72}}$

22. $\dfrac{a-b}{\sqrt{a}-\sqrt{b}}$

23. $\dfrac{2x+3y+5\sqrt{xy}}{\sqrt{x}+\sqrt{y}}$

24. $\dfrac{a+\sqrt{ab}}{b+\sqrt{ab}}$

C

Find the numerical values of the following:

1. $9^{\frac{1}{2}}\cdot 3^{-1}$

2. $8^{\frac{2}{3}}\cdot(4x)^0$

3. $\left(\frac{2}{3}\right)^{-2}+\left(\frac{1}{64}\right)^{\frac{1}{3}}$

4. $(2x^3)^0-(-2x^0)^3$

5. $\left(-\frac{1}{8}\right)^{-\frac{2}{3}}$

6. $2^0+2^{-1}+2^{-2}$

7. $27^{-\frac{2}{3}}\cdot 9^{\frac{3}{2}}$

8. $2^{-35}\cdot 2^{37}$

9. $\left(\frac{1}{4}\right)^{\frac{1}{2}}+(4)^{-\frac{1}{2}}$

10. $16^{\frac{3}{4}}\left(8^{-\frac{2}{3}}+4^{-\frac{3}{2}}\right)$

11. $(1^0+2^0+3^0+4^0)^{-\frac{1}{2}}$

12. $(3^{-1}+3^{-2})^{\frac{1}{2}}$

Perform the indicated operations, using the laws of exponents:

13. $a^4\cdot a^{-1}\cdot a^{-2}$

14. $(x^{-2})^{-2}\cdot(x^2)^{-1}$

15. $5ax^{\frac{1}{2}}\cdot 7a^{\frac{1}{3}}x^{-\frac{1}{3}}$

16. $\left(xy^{\frac{1}{2}}\right)^3\cdot\left(x^{\frac{1}{2}}y\right)^4$

17. $\left(\frac{m^{-2}}{n}\right)^{-1}\cdot\left(\frac{m}{n^{-2}}\right)^2$

18. $\dfrac{9b^2x^{\frac{1}{2}}}{6bx^{-\frac{1}{2}}}$

19. $\dfrac{(ab^2)^{\frac{1}{2}}}{(a^2b^3)^{-1}}$ 20. $\dfrac{3ax^{-1}\cdot 4a^2x^{\frac{1}{2}}}{6a^{-1}x^{-2}}$ 21. $\dfrac{\left(2x^{\frac{1}{2}}y^{\frac{1}{3}}\right)^6}{8x^{-1}y}$

Write with positive exponents, and simplify:

22. $\left(a^{\frac{1}{2}}-a^{-\frac{1}{2}}\right)^2$

23. $(x^{-1}+y^{-1})(x+y)^{-1}$

24. $\left(\dfrac{m^{\frac{1}{2}}}{n^{\frac{1}{2}}}-\dfrac{n^{\frac{1}{2}}}{m^{\frac{1}{2}}}\right)\cdot\dfrac{1}{n^{-1}-m^{-1}}$

25. $(1-x^{-1})^{-1}$

26. $\dfrac{ab^{-1}-a^{-1}b}{a^{-1}-b^{-1}}$

27. $\dfrac{x-y^{-1}z^2}{xyz^{-1}-z}$

Perform the indicated operations, and simplify:

28. $a^n\cdot a^{-\frac{n}{2}}\cdot a^{1-\frac{n}{2}}$ 29. $x^{2n-1}y^n\cdot x^{3-n}y^{1-n}$ 30. $(a^kb^2)^{-3}\cdot(a^{2-2k}b^{2k})^{\frac{1}{2}}$

D (optional)

Multiply:

1. $\sqrt[3]{9}\cdot\sqrt{6}$

2. $\sqrt{\frac{3}{2}}\cdot\sqrt[3]{\frac{4}{3}}$

3. $\sqrt[3]{9}\cdot\sqrt[4]{27}$

4. $\sqrt{18}\cdot\sqrt[4]{8}$

5. $\sqrt{\frac{1}{2}}\cdot\sqrt[3]{\frac{3}{4}}\cdot\sqrt[6]{\frac{8}{9}}$

6. $\sqrt{\frac{1}{2}}\cdot\sqrt[4]{\frac{2}{3}}\cdot\sqrt[8]{\frac{9}{64}}$

7. $(\sqrt{2}+\sqrt[3]{2})(\sqrt{2}-\sqrt[3]{2})$

8. $(\sqrt{2}+\sqrt[4]{6})(\sqrt{3}-\sqrt[4]{6})$

9. $\sqrt[4]{a^3b^2}\cdot\sqrt[5]{a^3b^4}$

10. $\sqrt[3]{4x^2y^2}\cdot\sqrt[4]{8x^3y^3}$

11. $\sqrt[3]{9m^2n^2}\div\sqrt{3mn}$

12. $2a^2b\sqrt[3]{3a^2b}\div 3b\sqrt[4]{9a^3b}$

E

Solve:

1. $x+\sqrt{x^2+7}=1$

2. $4+\sqrt{2x-1}=0$

3. $2+\sqrt[3]{x-4}=0$

4. $\sqrt[3]{x^3+3x^2}=x+1$

5. $1+\sqrt{x}=\sqrt{x+3}$

6. $2x-\sqrt{4x^2-15}=1$

7. $\sqrt{4x+1}=3-2\sqrt{x}$

8. $\sqrt{x-4}+\sqrt{x+4}=2$

9. $\sqrt{9x+40} = 3\sqrt{x} + 2$

10. $2 + \sqrt{x+3} = 3 + \sqrt{x+2}$

11. $\sqrt{x+5} - 1 = \frac{x}{\sqrt{x+5}}$

12. $\sqrt{x-3} + \sqrt{x} = \frac{6}{\sqrt{x}}$

F

1. Find the hypotenuse of a right triangle whose legs are of length 28 and 45 respectively.

2. Find the smallest whole number which must he added to 4731 to make a square number.

3. The area of a sphere is given by the formula $S = 4\pi r^2$. Find the radius of a sphere whose area is 20 square cm.

4. A city park is a half mile long and 350 yards wide (look up on the internet for how many yards there are to a mile). How much distance is saved by walking diagonally through the park rather than along two sides?

5. A ladder 20 feet long just reaches a window when the base is 4 feet from the wall. How high above the ground is the window?

6. Two buildings are 22 feet apart. A ladder 25 feet long is placed with its base on the ground between the tow buildings. If it reaches a window 20 feet from the ground on one building, how high will it reach on the other building if its base remains at the same point?

7. A swimming pool is 100 yards long and 160 feet wide. How far is it from one corner to the opposite corner (Hint: be careful with the units)?

8. In the formula $V = \pi r^2 h$, what is the value of r when $h = 6$ and $V = 231$?

9. A boat leaves a dock and travels eastward for 38 kilometers, southward for 36 kilometers, and then eastward for 39 kilometers more. How far is the boat from the dock?

10. A ladder 26 feet long is placed against a wall and it reaches a point 24 feet from the ground. If its base were placed only half as far from the wall, how much higher would the ladder reach up the wall?

Candlelight

Although a certain three candles are of the same length, it is known that they will burn for different lengths of time. After being lighted, it is observed, at intervals of $2, 2\frac{1}{2}$, and 3 hours, that the remainder of one candle is twice as long as the remainder of one of the others, How many hours will each candle burn?

A Square Trick

Find the simplified numerical value of

$$\sqrt{7+\sqrt{48}}+\sqrt{\sqrt{7}-\sqrt{48}}$$

Archimedes' Principle

If 19 grams of gold weigh but 18 grams when submerged in water, and 10 grams of silver then weigh 9 grams. how many ounces of silver and gold are there is a mass of an alloy of the two metals which weighs 387 grams in air and 351 grams in water?

Squares Squared

Show that the product $(a^2+b^2)(c^2+d^2)$ can be written as the sum of two squares.

Summary Sheets

Exponent laws:

Law	Example
$x^1 = x$	$4^1 = 4$
$x^0 = 1$	$5^0 = 1$
$x^m \cdot x^n = x^{m+n}$	$3^4 \cdot 3^5 = 3^9$
$\frac{x^m}{x^n} = x^{m-n}$	$\frac{4^8}{4^3} = 4^5$
$(x^m)^n = x^{mn}$	$(3^2)^4 = 3^8$
$(xy)^m = x^m y^m$	$(3 \cdot 2)^3 = 3^3 \cdot 2^3$
$\left(\frac{x}{y}\right)^m = \frac{x^m}{y^m}$	$\left(\frac{2}{3}\right)^3 = \frac{2^3}{3^3}$
$x^{-m} = \frac{1}{x^m}$	$4^{-3} = \frac{1}{4^3}$

Radical laws:

Law	Example
$\sqrt[n]{ab} = \sqrt[n]{a}\,\sqrt[n]{b}$	$\sqrt{4 \cdot 9} = \sqrt{4} \cdot \sqrt{9}$
$\sqrt[n]{\dfrac{a}{b}} = \dfrac{\sqrt[n]{a}}{\sqrt[n]{b}}$	$\sqrt{\dfrac{4}{9}} = \dfrac{\sqrt{4}}{\sqrt{9}}$
$\left(\sqrt[n]{a}\right)^n = a$	$\left(\sqrt[4]{25}\right)^4 = 25$
$\left(\sqrt[n]{a^m}\right) = \left(\sqrt[n]{a}\right)^m$	$\left(\sqrt[3]{27^6}\right) = \left(\sqrt[3]{27}\right)^6$
$\sqrt[m]{\sqrt[n]{a}} = \sqrt[mn]{a}$	$\sqrt{\sqrt[3]{4}} = \sqrt[6]{4}$

Answers to all Exercises

Exercises 1

1. x^7	**2.** m^8	**3.** a^6	**4.** 3^7
5. y^{23}	**6.** -128	**7.** a^5x^7	**8.** $a^3b^3c^4$
9. x^6y^4	**10.** m^5n^5	**11.** x^3	**12.** $\dfrac{1}{y^2}$
13. 1	**14.** x^2y^2	**15.** a/b	**16.** $\dfrac{1}{x^2y}$
17. a	**18.** 1	**19.** x^2/z^2	**20.** $\dfrac{c^2}{b}$
21. x^{18}	**22.** x^4	**23.** $-b^{16}$	**24.** a^{25}
25. $-m^6$	**26.** y^{24}	**27.** x^6y^6	**28.** $a^{15}x^{16}$

29. $a^4b^9c^{16}$ **30.** $a^8x^9y^8$ **31.** $8x^9$ **32.** $9a^2$

33. $-a^8b^{12}$ **34.** $-x^{15}y^{12}$ **35.** $8a^3b^6$ **36.** a^2x^6

37. $a^8b^{12}c^4$ **38.** $x^{21}y^{14}z^{28}$ **39.** $16a^4x^8y^{12}$ **40.** $-27m^6n^3x^3$

41. $\frac{27}{64}$ **42.** $\frac{x^2}{16}$ **43.** $\frac{a^4}{x^8}$ **44.** $\frac{125}{p^3}$

45. $\frac{a^{15}}{b^{25}}$ **46.** $\frac{8x^3}{27y^6}$ **47.** $\frac{25a^6}{36b^4}$ **48.** $\frac{a^5b^{10}}{32x^{15}}$

49. $\frac{a^{10}b^{20}c^{10}}{x^{20}y^{30}z^{10}}$ **50.** $\frac{x^{16}y^{24}}{256a^{16}b^8}$

51. a^5x^{13} **52.** $8m^8n^5$ **53.** $\frac{y^5}{x^{10}}$ **54.** b^5c^{13}

55. $x^{12}y^8$ **56.** $a^{10}b^{11}$ **57.** $72a^3b^2x^{12}$ **58.** $25a^4x^6b^9y^{12}$

59. a **60.** $\frac{3p^4}{4q}$ **61.** $\frac{y^{10}}{x^3}$ **62.** $\frac{3a^7b}{2}$

63. $\frac{1}{n^7}$ **64.** $\frac{a^2x^5}{2}$ **65.** $\frac{x^4y}{ab^4}$ **66.** $\frac{x^{13}}{y^4}$

67. $a^4b^3x^4y^3$ **68.** ax^4y^9 **69.** $6c$ **70.** $\frac{x^2}{a^2}$

Exercises 2

1. ± 4 **2.** ± 6 **3.** ± 8 **4.** ± 9 **5.** $\pm a$

6. $\pm b^2$ **7.** $\pm x^2y$ **8.** $\pm 7m$ **9.** $\pm 5a^2$ **10.** $\pm 20c^5$

11. 2 **12.** 3 **13.** -4 **14.** 5 **15.** a

16. $-c^3$ **17.** $4xy$ **18.** $10m^2$ **19.** $-a^2b$ **20.** $2/3a$

21. 11 **22.** -12 **23.** 3 **24.** 15 **25.** 2

26. 9 **27.** -2 **28.** -25 **29.** $\frac{1}{4}$ **30.** 2

31. b^2 **32.** $4mn$ **33.** 4 **34.** -1 **35.** $3a$

36. $4a$ **37.** 10 **38.** 2 **39.** $11a^2$ **40.** ab

41. $\frac{1}{6}$

Exercises 3

1. $\sqrt{x}$, $\sqrt[3]{x}$, $\sqrt[3]{y^2}$, $\sqrt[4]{y^3}$, $\sqrt[7]{a}$, $\sqrt[5]{a^3}$, $\sqrt{b}$, $\sqrt[5]{b^2}$

2. $\sqrt{x^3}$, $\sqrt[3]{x^4}$, $\sqrt[6]{y^5}$, $\sqrt[5]{y^6}$, $\sqrt[2]{a^5}$, $\sqrt[5]{a^2}$, $\sqrt{a+b}$, $\sqrt[3]{a-b}$

3. $3\sqrt{x}$, $\sqrt{3x}$, $8\sqrt[3]{y}$, $5\sqrt[5]{y^3}$, $2\sqrt[3]{a^4}$, $2\sqrt[4]{a^3}$, $\sqrt{a^2+b^2}$, $\sqrt[3]{a^3-b^3}$

4. $4\sqrt[3]{x^4}$, $2\sqrt{x^7}$, $\sqrt[3]{16y^2}$, $4\sqrt[3]{y^2}$, $3\sqrt{a^3}$, $\sqrt[3]{16a^4}$, $7\sqrt{a+b}$, $\sqrt{7(a+b)}$

5. $a^{\frac{1}{2}}$, $a^{\frac{1}{3}}$, $b^{\frac{2}{3}}$, $b^{\frac{3}{2}}$, $x^{\frac{2}{5}}$, $x^{\frac{5}{2}}$, $y^{\frac{3}{4}}$, $y^{\frac{5}{3}}$

6. $a^{4/3}$, $a^{1/4}$, $b^{2/6}$, $b^{7/3}$, $x^{2/9}$, $x^{7/4}$, $(x+y)^{1/2}$, $(x^2+y^2)^{1/2}$

7. $2a^{\frac{1}{2}}$, $(2a)^{\frac{1}{2}}$, $7b^{\frac{2}{3}}$, $3b^{\frac{3}{5}}$, $2x^{\frac{3}{4}}$, $5x^{\frac{5}{2}}$, $2(x-y)^{\frac{1}{2}}$, $4\left(x^2-y^2\right)^{\frac{1}{2}}$

8. $ab^{1/2}$,$(ab)^{1/2}$,$ba^{1/2}$,$ab^{2/3}$, $(ab^2)^{1/3}$, $ba^{1/3}$, $x(x+y)^{1/2}$,$x(x-y)^{1/3}$

9. 5	**10.** 2	**11.** 2	**12.** 9	**13.** 8
14. 4	**15.** 12	**16.** 2	**17.** −1	**18.** 1
19. 27	**20.** 27	**21.** $\frac{1}{2}$	**22.** 5/6	**23.** $\frac{4}{9}$
24. 8/27	**25.** 0.5	**26.** 1.1	**27.** 0.3	**28.** 0.25
29. $a^{\frac{3}{4}}$	**30.** $x^{\frac{2}{3}}$	**31.** $b^{\frac{3}{2}}$	**32.** $b^{3/2}$	**33.** 1
34. $x^{13/6}$	**35.** n	**36.** a^2	**37.** $a^{\frac{1}{2}}$	**38.** $\sqrt{x}$
39. $y^{\frac{5}{12}}$	**40.** 2	**41.** $n^{\frac{7}{12}}$	**42.** $\sqrt[4]{x}$	**43.** $x^{\frac{1}{6}}$
44. $m^{7/6}$	**45.** a^4	**46.** a^4	**47.** $x^{\frac{1}{3}}$	**48.** y
49. 2	**50.** z^4	**51.** x^3y^2	**52.** $\sqrt[3]{a}\sqrt{b}$	**53.** $x^{\frac{5}{3}}/y^{\frac{5}{2}}$
54. a^6/b^4	**55.** $b^{\frac{9}{2}}/c^4$	**56.** $x^{3/2}/y^6$	**57.** $x^{\frac{5}{2}}$	**58.** ax
59. $n^{\frac{10}{3}}$	**60.** 1	**61.** $a^{\frac{1}{2}}b$	**62.** $\sqrt[6]{a}\sqrt{x}$	

Exercises 4

1. $\frac{1}{4}$	**2.** 1	**3.** $\frac{1}{8}$
4. 0.01	**5.** 1	**6.** 9/4
7. 5	**8.** 1	**9.** $\frac{1}{16}$
10. 125/8	**11.** $\frac{1}{2}$	**12.** 3
13. $\frac{8}{9}$	**14.** 32	**15.** $\frac{3}{10000}$
16. 1/2	**17.** 2	**18.** 18
19. $-\frac{1}{4}$	**20.** −1/2	**21.** x
22. $a^{3/2}$	**23.** m^{-4}	**24.** b^4
25. x^4	**26.** y^7	**27.** x^6
28. $1/a^8$	**29.** y^3	**30.** x^2y^4
31. ab^{-4}	**32.** x^4/a^{12}	**33.** $1/a^4$
34. b^3/a^2	**35.** $1/5x^2$	**36.** $3/y^3$
37. x^2/a^3	**38.** $c^3/(ab^2)$	**39.** $2y/3x^4$

40. $a^2y^3/(x^2b^3)$	**41.** $(y-x)/xy$	**42.** $ab+ba$
43. $1/(x-y)$	**44.** $8m^3n^2$	**45.** $2x/y$
46. $z/x+z/y$	**47.** $(a^2+b^2)/ab$	**48.** $\sqrt{1/(x+y)^2}$
49. $(n^2-m^2)/m^2n^2$	**50.** $\dfrac{a^2+b^2}{a^2-b^2}$	**51.** $xy(x-y)$
52. $ab/(a+b)^2$	**53.** a^3	**54.** m^{2m-2}
55. b^{2-n}	**56.** y^{3-n}	**57.** x^{n^2-n}/y^{3n-3}
58. a^3/b	**59.** $a^{2n}b^{n^2}$	**60.** $(xy)^{2a}/y^2$
61. a^{k+2}	**62.** $xy^{1-n/2}$	**63.** p^rq^{-r}
64. $m^{2-x}n^{4x-3}$		

Exercises 5

1. 4	**2.** 6	**3.** 5	**4.** 3
5. -2	**6.** 2	**7.** $2a$	**8.** $3x^2$
9. $-3b^2$	**10.** $2\sqrt{3}$	**11.** $5\sqrt{2}$	**12.** $6\sqrt{2}$
13. $11\sqrt{2}$	**14.** $5\sqrt{3}$	**15.** $9\sqrt{2}$	**16.** $7\sqrt{3}$
17. $20\sqrt{2}$	**18.** $2\sqrt[3]{3}$	**19.** $3\sqrt[3]{2}$	**20.** $2\sqrt[3]{-5}$
21. $-5\sqrt[3]{2}$	**22.** $2\sqrt[4]{2}$	**23.** $2\sqrt[4]{3}$	**24.** $2\sqrt{5}\sqrt{x^3}$
25. $3ab\sqrt{5a}$	**26.** $5\sqrt{7}\sqrt{xy^2}$	**27.** $3a\sqrt{7}$	**28.** $2\sqrt[3]{6}\sqrt[3]{x^5}$
29. $3ab^3\sqrt[3]{3a^2}$	**30.** $2\sqrt{5}$	**31.** $7\sqrt{3}$	**32.** $3\sqrt{10}$
33. $15\sqrt{7}$	**34.** $7\sqrt{66}$	**35.** $5\sqrt{42}$	**36.** $3\sqrt[3]{2}$
37. $2\sqrt[3]{7}$	**38.** $30\sqrt{6}$	**39.** $20a^3$	**40.** $56\cdot 3^{2/3}$

Exercises 6

1. 2	**2.** 6	**3.** 7	**4.** 6
5. 2	**6.** 3	**7.** -2	**8.** 2
9. $2a$	**10.** $5x^2y^3$	**11.** x^2	**12.** a^2b
13. $\frac{1}{5}\sqrt{15}$	**14.** $\sqrt{35}/7$	**15.** $\frac{1}{6}\sqrt{15}$	**16.** $\frac{1}{6}\sqrt{14}$
17. $\frac{1}{2}\sqrt{10}$	**18.** $\frac{4}{3}\sqrt{2}$	**19.** $\frac{3}{5}\sqrt{15}$	**20.** $\frac{7}{6}\sqrt{6}$
21. $\frac{1}{2}\sqrt[3]{6}$	**22.** $4\sqrt[3]{3}/3$	**23.** $\frac{1}{2b^2}\sqrt{6ab}$	**24.** $\frac{2\sqrt{35}}{15\sqrt{n}}$
25. $\sqrt[3]{xy^2}/y$	**26.** $\sqrt[3]{\frac{2}{3}}\sqrt[3]{\frac{a^2}{b^2}}$	**27.** $\sqrt{2ab}$	**28.** $2\sqrt[3]{18}$
29. $1/10\sqrt{70}$	**30.** $5\sqrt{6}/4$	**31.** $\frac{2}{5}\sqrt{35}$	**32.** $\frac{1}{2}\sqrt{3}$
33. $\frac{1}{25}\sqrt{15}$	**34.** $\frac{1}{4}\sqrt{10}$	**35.** $\frac{1}{6}\sqrt{6}$	**36.** $\frac{1}{3}\sqrt{10}$
37. $\frac{1}{2}\sqrt[3]{15}$	**38.** $5\sqrt[3]{3}/3$	**39.** $\sqrt[3]{4}$	**40.** 3/5

41. $\sqrt{5b}$	**42.** $\frac{\sqrt[3]{5}a^{2/3}}{3\cdot 2^{2/3}}$	**43.** $\frac{1}{2}x\sqrt{2y}$	**44.** $\frac{a}{b}$

Exercises 7

1. $7\sqrt{2}$	**2.** $7\sqrt{3}$	**3.** $\sqrt{5}$	**4.** $\sqrt{2}$
5. $3\sqrt{3}$	**6.** $7\sqrt{5}$	**7.** $11\sqrt{2}$	**8.** $2\sqrt{3}$
9. $3\sqrt{2}$	**10.** $11\sqrt{5}$	**11.** $4\sqrt{5}$	**12** $4\sqrt{3}$
13. $18\sqrt{7}$	**14.** $-5\sqrt{10}$	**15.** $12\sqrt{2}$	**16** $-\sqrt{2}$
17. $-4\sqrt{5}$	**18.** $10\sqrt{2}$	**19.** $3\sqrt[3]{3}$	**20** $\sqrt[3]{2}$
21. $3\sqrt[3]{5}$	**22** $\sqrt[3]{7}$	**23.** $13\sqrt[3]{6}$	**24.** $2\sqrt[3]{2}$
25. $2\sqrt[3]{3}$	**26.** $11\sqrt{5}-5$	**27.** $2\sqrt{7}+6\sqrt{5}$	**28.** $26\sqrt{2}+2\sqrt{3}$
29. $67\sqrt{3}-42\sqrt{5}$	**30.** $\frac{1}{2}(\sqrt{2}+\sqrt{6})$	**31.** $\sqrt{3}+2\sqrt{15}$	**32.** $\sqrt{3}$
33. 0	**34.** $13\sqrt[3]{2}+6\sqrt[3]{2}$	**35.** $8-3\sqrt[3]{6}$	**36.** $7+2\sqrt[3]{5}$

Exercises 8

1. $3\sqrt{2}+6$	**2.** $5(1+\sqrt{3})$
3. $2+\sqrt{6}-2\sqrt{3}$	**4.** $2\sqrt{15}-4\sqrt{3}-3\sqrt{14}$
5. $6\sqrt{3}-6\sqrt{2}-6$	**6.** $10\sqrt{2}-10\sqrt{5}+15\sqrt{6}$
7. $40-48\sqrt{3}-16\sqrt{5}$	**8.** $24\sqrt{3}-45\sqrt{2}+54$
9. $108-40\sqrt{6}-24\sqrt{5}$	**10.** $210+90\sqrt{5}+60\sqrt{6}$
11. 1	**12.** $7+2\sqrt{10}$
13. $12-11\sqrt{21}$	**14.** $23-43\sqrt{15}$
15. $35+12\sqrt{6}$	**16.** $6+2\sqrt{6}$
17. $14\sqrt{6}-4$	**18.** $372-176\sqrt{6}$
19. $756-216\sqrt{6}$	**20.** $400\sqrt{6}+1100$
21. $1-2\sqrt[3]{6}+\sqrt[3]{36}$	**22.** 3
23. $132\sqrt{3}-162\sqrt{2}$	**24.** $31+10\sqrt{6}+6\sqrt{10}+4\sqrt{15}$
25. $x(\sqrt{a}-\sqrt{b}+\sqrt{x})$	**26.** $x-y$
27. $a+b-2\sqrt{ab}$	**28.** $a\sqrt{b}+\sqrt{a}b$
29. $1-2\sqrt{n(1-n)}$	**30.** $2\sqrt{x-y}\sqrt{x+y}+2x$
31. $3a-8+a\sqrt{a-3}$	**32.** $1-2x\sqrt{2-4x^2}$
33. $9+5\sqrt{2}$	**34.** $\sqrt{3}-6$
35. 0	**36.** 0
37. $17+\sqrt{2}-\sqrt{3}-4\sqrt{6}$	**38.** $4(19+8\sqrt{5})$

Exercises 9

1. $\sqrt{2}-3\sqrt{5}+4\sqrt{7}$

2. $3\sqrt{3}-5\sqrt{11}-2\sqrt{5}$

3. $4-3\sqrt{6}+\sqrt{10}$

4. $2\sqrt{15}+3\sqrt{21}+6\sqrt{33}$

5. $\frac{5}{6}\sqrt{6}+\frac{10}{3}\sqrt{3}-\frac{15}{2}\sqrt{2}$

6. $\sqrt{21}-4\sqrt{35}-3\sqrt{91}$

7. $\frac{1}{4}\sqrt{30}+\frac{1}{3}\sqrt{78}-\frac{1}{6}\sqrt{114}$

8. $\frac{\sqrt{2}}{4}-\frac{\sqrt{35}}{4}-\frac{3\sqrt{3}}{4}$

9. $\sqrt{3}+1$

10. $\sqrt{5}+3$

11. $3(\sqrt{3}-\sqrt{2})$

12. $2\sqrt{2}-\sqrt{3}$

13. $\sqrt{5}-2\sqrt{2}$

14. $2+\sqrt{7}$

15. $9+4\sqrt{5}$

16. $6-\sqrt{35}$

17. $\sqrt{6}+2$

18. $3\sqrt{5}-2\sqrt{7}$

19. $(\sqrt{a}-\sqrt{b})/(a-b)$

20. $(\sqrt{x}+\sqrt{y})^2/(x-y)$

21. $\sqrt{15}$

22. $3-2\sqrt{2}$

23. $6+2\sqrt{2}$

24. $28-2\sqrt{6}$

25. $\sqrt{2}$

26. $2\sqrt{2}$

Exercises 10

1. $\sqrt{2}$

2. $\sqrt[3]{5}$

3. $\sqrt{2}$

4. $\sqrt[4]{7}$

5. $\sqrt[3]{9}$

6. $\sqrt[3]{4}$

7. $\sqrt{2}$

8. $\sqrt[4]{5}$

9. $\sqrt{2}$

10. $\sqrt[3]{2}\sqrt[6]{3}$

11. $\sqrt[5]{11}$

12. $\sqrt{3}$

13. $\sqrt{3xy}$

14. $\sqrt{2a}$

15. $\sqrt{2mn}$

16. $\sqrt{2}\sqrt{b}\sqrt[4]{ac^3}$

17. $\sqrt[6]{27}, \sqrt[6]{25}$

18. $\sqrt[8]{49}, \sqrt[8]{6561}$

19. $\sqrt[20]{32}, \sqrt[20]{81}$

20. $\sqrt[12]{256}, \sqrt[12]{81}$

21. $\sqrt[6]{49}, \sqrt[6]{42}$

22. $\sqrt[15]{243}, \sqrt[15]{125},$

23. $\sqrt[8]{16}, \sqrt[8]{16}, \sqrt[8]{8}$

24. $\sqrt[12]{16}, \sqrt[12]{27}, \sqrt[12]{25}$

25. $\sqrt[12]{729}, \sqrt[12]{625}, \sqrt[12]{343}$

26. $\sqrt[6]{8a^3}, \sqrt[6]{9b^4}$

27. $\sqrt[12]{x^9}, \sqrt[12]{9x^{10}}$

28. $\sqrt[12]{81m^4}, \sqrt[12]{8n^9}$

29. $\sqrt[6]{1125}$

30. $\sqrt[12]{432}$

31. $\sqrt[4]{539}$

32. $\sqrt[12]{576}$

33. $\sqrt[6]{40}$

34. $\sqrt[10]{224}$

35. $\sqrt[10]{a^{13}b^7}$

36. $\sqrt[21]{x^{29}}$

37. $mn\sqrt[12]{40m^{12}}$

38. $\sqrt[6]{2187a^7}$

39. $ab\sqrt[30]{a^9b^{13}}$

40. $xy\sqrt[10]{63}$

Exercises 11

1. $2i$

2. $5i$

3. $8i$

4. $2i\sqrt{2}$

5. $2\sqrt{3}i$

6. $6i$

7. $21i$

8. $10i\sqrt{2}$

9. $15\sqrt{2}i$	**10.** $12i\sqrt{5}$	**11.** $2ai$	**12.** $4iax\sqrt{2}$
13. $2x^2\sqrt{5}i$	**14.** $3ix\sqrt{x}$	**15.** $2a^2\sqrt{2ai}$	**16.** $-i$
17. 1	**18.** i	**19.** i	**20.** i
21. -1	**22.** $-i$	**23.** -1	**24.** i
25. 1	**26.** i	**27.** 1	**28.** -1
29. $-i$	**30.** -1		

Exercises 12

1. $x = 13$	**2.** 13	**3.** No root	**4.** 1
5. No root	**6.** No root	**7.** $x = 2$	**8.** 4
9. $x = -29$	**10.** 65/2	**11.** $x = 0$	**12.** No root
13. No root	**14.** 5	**15.** $x = 5$	**16.** -4
17. $x = 4$	**18.** 5	**19.** $x = \frac{5}{2}$	**20.** No root
21. No root	**22.** No root	**23.** $x = 50$	**24.** 37

Exercises 13

1. 2.98	**2.** 889	**3.** 70.8	**4.** 17.3
5. 0.939	**6.** 8.72	**7.** 109	**8.** 2.23
9. 0.497	**10.** 0.171	**11.** 0.022	**12.** 60.6
13. 4.13	**14.** 1.37	**15.** 4.89	**16.** 3.07
17. 0.89	**18.** 8.24	**19.** 0.23	**20.** 2.24
21. 4.47	**22.** 3.51	**23.** 4.08	**24.** 15.91

Exercise 14

1. 4.359	**2.** 9.695	**3.** 8.124	**4.** 6.403
5. 9.327	**6.** 23	**7.** 96	**8.** 58
9. 44	**10.** 76	**11.** 15.10	**12.** 19.21
13. 13.23	**14.** 12.12	**15.** 24.60	**16.** 1.84
17. 1.659	**18.** 4.54	**19.** 5.888	**20.** 3.376
21. 1.342	**22.** 2.99	**23.** 1.803	**24.** 1.72
25. .7798	**26.** 7.035	**27.** 6.854	**28.** 1.828
29. 4.560	**30.** 1.082		

Exercise 15

1. $c = 10$	**2.** $b = 16$	**3.** $a = 24$	**4.** $c = 29$
5. $a = 27$	**6.** $b = 75$	**7.** 27.9	**8.** 11.2

9. 111.0 ft. **10.** 78.48. **11.** 105.3 ft. **12.** $2\sqrt{m^2n^2}$.

Answers to Review Questions

A.

1. $11\sqrt{2}$ **2.** $6\sqrt{3}$ **3.** $2a\sqrt{3a}$

4. $60\sqrt{2}$ **5.** $7(x-y)\sqrt{2}$ **6.** $18\sqrt{5}m^2n\sqrt{m^5n^3}$

7. $(a+b)\sqrt{a-b}$ **8.** $-10\sqrt[3]{12}$ **9.** $4ab\sqrt[3]{a^2c^2}$

10. $\frac{\sqrt{7}}{3\sqrt{3}}\sqrt{\frac{a^3}{d^5}}$ **11.** $8\sqrt{3}$ **12.** $2\frac{\sqrt{(ax)/(by))}}{3\sqrt{3}}$

13. $\frac{4}{7}\sqrt{3}$ **14.** $\frac{3}{\sqrt[3]{2}}$ **15.** $\frac{2}{\sqrt[3]{3}}\sqrt[3]{\frac{a^4}{x^2}}$

16. $9\sqrt{2}$ **17.** $6x\sqrt{2}$ **18.** $\sqrt{6}$

19. $\sqrt[3]{2}$ **20.** $\frac{1}{3\sqrt[3]{2}}$ or $\frac{\sqrt[3]{4}}{6}$ **21.** $(a-b)\sqrt{2}$

22. $\frac{1}{10}(5\sqrt{6}-7\sqrt{5})$ **23.** $11\sqrt{5}-39\sqrt{2}$ **24.** $2\sqrt{2}+4\sqrt{3}$

25. $\frac{5}{\sqrt[3]{6}}+2\sqrt[3]{2}$ **26.** $\sqrt{b}(a-b+c)$ **27.** $(a-b)\sqrt{a-b}$

28. $\sqrt{x}y(x^2y^2+xy+1)$ **29.** $(p+q)^2\sqrt[3]{p}$ **30.** $(x(y+z)+yz)/\sqrt{xyz}$

31. $\sqrt{a/b}$ **32.** $\sqrt{a(x-1)^2}+\sqrt{a(x+1)^2}$

33. $2\sqrt{x-2}$

B.

1. 9 **2.** 270 **3.** 120

4. 1/15 **5.** 3780 **6.** $\sqrt{6}/30$

7. $12\sqrt{5}-36+24\sqrt{7}$ **8.** $\sqrt{a}bc+ab\sqrt{c}+a\sqrt{b}c$ **9.** $17+4\sqrt{10}$

10. $85-60\sqrt{2}$ **11.** $\sqrt[3]{4}-2\sqrt[3]{2}$ **12.** 11

13. $2\sqrt{6}$ **14.** $5\sqrt{3}$ **15.** $3\sqrt[3]{4}$

16. $\sqrt[3]{5}$ **17.** $4\sqrt{7}-\sqrt{3}-4$ **18.** $\sqrt{6}/6+\sqrt{5}$

19. $2(\sqrt{7}+2)$ **20.** $4\sqrt{10}+3\sqrt{15}$ **21.** $3+\sqrt{6}$

22. $\sqrt{a}+\sqrt{b}$ **23.** $2\sqrt{x}+3\sqrt{y}$ **24.** $\sqrt{\frac{a}{b}}$

C.

1. 1 **2.** 4 **3.** $2\frac{1}{2}$ **4.** 9

5. 4	**6.** 7/4	**7.** 3	**8.** 4
9. 1	**10.** 3	**11.** $\frac{1}{2}$	**12.** 2/3
13. a	**14.** x^2	**15.** $35a^{\frac{4}{3}}x^{\frac{1}{6}}$	**16.** $x^5y^{11/2}$
17. m^4n^5	**18.** $3bx/2$	**19.** $a^{\frac{5}{2}}b^4$	**20.** $2a^4x^{3/2}$
21. $8x^4y$	**22.** $(a-1)^2/a$	**23.** $1/xy$	**24.** $\sqrt{m}\sqrt{n}$
25. $x/(x-1)$	**26.** $-a-b$	**27.** z/y	**28.** a
29. $x^{n+2}y$	**30.** $a^{1-4k}\,b^{k-6}$		

D.

1. $3\sqrt[6]{24}$ (see Note 1)	**2.** $\sqrt[6]{6}$	**3.** $3\sqrt[12]{243}$
4. $6\sqrt[4]{2}$	**5.** $\frac{1}{2}\sqrt[3]{2}$	**6.** $\sqrt[8]{\frac{1}{96}}$
7. $2-\sqrt[3]{4}$	**8.** $\sqrt[4]{6}(\sqrt{3}-\sqrt{2})$	**9.** $ab\sqrt[20]{a^7b^6}$
10. $\sqrt[3]{4}\sqrt[4]{8}\cdot\sqrt[12]{x^{17}y^{17}}$	**11.** $\sqrt[6]{3mn}$	**12.** $(2\sqrt[12]{a^{23}b})/(3\sqrt[6]{3})$

Note 1:

$$\sqrt[3]{9}\sqrt{6} = 9^{\frac{1}{3}}6^{\frac{1}{2}} = (3^2)^{\frac{1}{3}}6^{\frac{1}{2}} = 3^{\frac{2}{3}}6^{\frac{1}{2}} = 3^{\frac{2}{3}}(2\times 3)^{\frac{1}{2}} = 3^{\frac{2}{3}}3^{\frac{1}{2}}\,2^{\frac{1}{2}}$$

$$3^{\frac{2}{3}}3^{\frac{1}{2}}\,2^{\frac{1}{2}} = 3^{\frac{2}{3}+\frac{1}{2}}\,2^{\frac{1}{2}} = 3^{1+\frac{1}{6}}\,2^{\frac{1}{2}} = 3\cdot 3^{\frac{1}{6}}\,2^{\frac{3}{6}}$$

$$3\cdot(3^1\cdot 2^3)^{\frac{1}{6}} = 3\cdot(24)^{\frac{1}{6}} = 3\sqrt[6]{24}$$

E.

1. $x=-3$	**2.** No root	**3.** $x=-4$	**4.** $-1/3$
5. $x=1$	**6.** 4	**7.** $x=\frac{4}{9}$	**8.** No root
9. $x=9$	**10.** -2	**11.** $x=20$	**12.** 4

F.

1. 53	**2.** 30	**3.** 1.26 cms.	**4.** 282.95 yards
5. 19.6 ft.	**6.** 24 ft.	**7.** 340 ft.	**8.** 2.305
9. 85 kilometers	**10.** 1.515 ft.		

Made in the USA
Monee, IL
04 November 2026

67376426R00046